3, 0, 5, 9 → Pg 41 & 52

#3, #5

LABORATORY MANUAL
for PAYNTER'S

SECOND EDITION

Introductory
Electronic Devices
and Circuits

SPRING 92
17 & 22

EX 4 3 LABS #5
EX 10

ROBERT T. PAYNTER / WALTER R. MILLER
CLIVE D. MENEZES / JOHN DESSEL

LABORATORY MANUAL
for PAYNTER'S

SECOND EDITION

Introductory Electronic Devices and Circuits

Robert T. Paynter

St. Louis Community College at Forest Park

Prentice Hall,
Englewood Cliffs, New Jersey 07632

Editorial/production supervison and
   interior design: *Maureen Lopez*
Manufacturing buyers: *Mary McCartney/Ed O'Dougherty*
Acquitions editor: *Holly Hodder*
Supplements acquitions editor: *Judith Casillo*

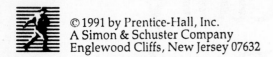

© 1991 by Prentice-Hall, Inc.
A Simon & Schuster Company
Englewood Cliffs, New Jersey 07632

Printed in the United States of America

10 9 8 7 6 5 4 3 2 1

0-13-483025-3

Prentice-Hall International (UK) Limited, *London*
Prentice-Hall of Australia Pty. Limited, *Sydney*
Prentice-Hall Canada Inc., *Toronto*
Prentice-Hall Hispanoamericana, S.A., *Mexico*
Prentice-Hall of India Private Limited, *New Delhi*
Prentice-Hall of Japan, Inc., *Tokyo*
Simon & Schuster Asia Pte. Ltd., *Singapore*
Editora Prentice-Hall do Brasil, Ltda., *Rio de Janeiro*

Contents

PART III   BJTs and BJT Circuits

PART IV   FETs and FET Circuits

PART V   Differential Amplifiers and Basic Op-Amp Circuits

PART VI   Amplifier Frequency Response

Preface

This lab manual has been written to accompany the second edition of *Introductory Electronic Devices and Circuits* by Robert T. Paynter. As such, the exercises have been arranged (as closely as possible) to follow the progression of topic coverage in the text.

The exercises in this manual have been tested (under supervision) in our labs by our students. At the same time, we recognize the fact that there is no such thing as a "universal" lab. That is, we realize that not every lab exercise will work the same way for every student in every school. Therefore, as a guideline, we have included the average results that our students obtained in the *Instructor's Resource Manual*. These results, while not absolute, will give you a good idea of the type of results that your students should obtain. We hope that this resource manual, along with the lab manual, make for a great learning experience in the lab.

Robert T. Paynter
Walter R. Miller
Clive Menezes
John Dessel

Parts List

RESISTORS

#	Value	#	Value	#	Value	#	Value	#	Value
1	10 Ω	2	100 Ω	2	1 kΩ	3	10 kΩ	2	100 kΩ
		1	120 Ω						
		1	220 Ω						
1	27 Ω	1	390 Ω	1	1.1 kΩ	1	11 kΩ	1	180 kΩ
		1	470 Ω						
1	33 Ω	1	510 Ω	1	1.2 kΩ	1	12 kΩ	1	200 kΩ
2	68 Ω	1	560 Ω	2	1.5 kΩ	4	15 kΩ	1	270 kΩ
		1	820 Ω						
1	91 Ω	1	910 Ω	1	1.6 kΩ	3	20 kΩ	1	470 kΩ
				1	2 kΩ	1	22 kΩ	1	1 MΩ
				2	2.2 kΩ	1	27 kΩ	1	1.5 MΩ
				2	3.3 kΩ	1	30 kΩ	1	15 MΩ
				2	3.6 kΩ	1	33 kΩ	1	22 MΩ
				1	3.9 kΩ	1	39 kΩ		
				1	5.1 kΩ	1	47 kΩ		
						1	51 kΩ		
				1	5.6 kΩ	1	82 kΩ		
				1	6.8 kΩ				
				1	8.2 kΩ				
				1	9.1 kΩ				

*Quantities (#) indicate the maximum number required in any single exercise.

CAPACITORS

#	Value	#	Value	#	Value	#	Value
1	22 pF	1	0.002 µF	3	1 µF	1	100 µF
1	51 pF	1	0.003 µF	2	4.7 µF	1	470 µF
1	100 pF	2	0.01µF	3	10 µF	1	1000 µF
1	470 pF	1	0.022 µF	1	22 µF	1	2200 µF
2	0.001 µF	1	0.1 µF	4	47 µF		
		1	0.47 µF				

POTENTIOMETERS

#	Value	#	Value	#	Value	#	Value
1	1 kΩ	1	10 kΩ	2	100 kΩ	1	2.5 MΩ
1	5 kΩ	1	25 kΩ	1	500 kΩ		
		2	50 kΩ				

DEVICES

#	Part Number	Description
2	1N3821	Zener diode
4	1N4001	Rectifier diode
2	1N4148	Small-signal diode
1	1N5240	Zener diode
1	2N2222	Npn transistor
3	2N3904	Npn transistor
2	2N3906	Pnp transistor
1	2N4444	Silicon controlled rectifier (SCR)
1	2N4870	Unijunction transistor (UJT)
2	2N5485	N-channel JFET
1	4N35	Opto-isolator
1	LM555	Timer
1	MLED71	Infrared LED
1	μA741 (or equivalent)	Operational amplifier
1	XXXX	LED, general-purpose, any color

MISCELLANEOUS

#	Item
1	24 Vac transformer (center-tapped)
1	Soldering iron
1	8 Ω speaker
1	1 mH inductor
1	10 mH inductor
1	Momentary push-button switch
1	2.5 cm (1 inch) piece of heat shrink wrap

OPTIONAL PARTS

#	Item	Description
1	TL071	Operational amplifier
4	μA741 (or equivalent)	Operational amplifiers (additional)

> Note: The optional parts listed above are used in THE BRAIN DRAIN portions of two of the exercises. They are not needed for the required portions of those exercises.

LABORATORY MANUAL
for PAYNTER'S

SECOND EDITION

Introductory Electronic Devices and Circuits

PART I

Test Equipment

Exercise 1

Meter Input Impedance

OBJECTIVES

- To demonstrate one method for measuring meter input impedance.
- To demonstrate the effects of meter input impedance on voltage measurements.

DISCUSSION

When a VOM or DMM is used to measure the voltage across a component, the meter is connected *across* the component. This creates a parallel circuit consisting of the component and the input impedance of the meter, as is shown in Figure 1.1. As long as the meter input impedance is *at least ten times* the value of R_2, the *measured* voltage across R_2, $V_{R2(M)}$, will be within ten percent of the actual voltage across the component, V_{R2}. However, if the input impedance of the meter is less than ten times the value of R_2, $V_{R2(M)}$ will be out of the ten percent tolerance that is normally considered to be acceptable. Therefore, it is important that you know the approximate value of Z_{in} for a given meter before using it for any voltage measurements.

VOMs typically have input impedance values that are in the neighborhood of *20 kΩ/V*. This means that the input impedance of the meter will vary from one voltage scale to another. For example, let's say that a meter has a 20 kΩ/V input impedance rating and dc voltage scales of 1 V, 10 V, and 100 V. The value of Z_{in} for the meter for each voltage scale would be found as

$$Z_{in} = 20 \text{ k}\Omega/V \times \quad 1V = 20 \text{ k}\Omega \text{ (when the 1 V scale is used)}$$

$$Z_{in} = 20 \text{ k}\Omega/V \times \quad 10V = 200 \text{ k}\Omega \text{ (when the 10 V scale is used)}$$

$$Z_{in} = 20 \text{ k}\Omega/V \times 100V = 2 \text{ M}\Omega \text{ (when the 100 V scale is used)}$$

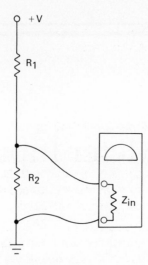

Figure 1.1

As you can see, the input impedance of the VOM changes as the voltage scale is changed.

DMMs normally have a single input impedance rating. In other words, the input impedance of the typical DMM will not change from one voltage scale to another. This Z_{in} rating is typically much higher than that of the VOM. The high Z_{in} of the DMM reduces the load effect on the circuit under test. This is why voltage readings made with a DMM tend to be more accurate than those made with a standard VOM.

LAB PREPARATION

Review Appendix C of *Introductory Electronic Devices and Circuits*.

LAB OVERVIEW

In this exercise, you will:

1. Determine the input impedance ratings of a VOM and a DMM using the documentation on the meters.
2. Measure the input impedance of your VOM.
3. Take a series of circuit voltage measurements using the VOM.
4. Repeat the circuit voltage measurements using the DMM.
5. Calculate the *percent of error* caused by using the VOM and the DMM for voltage measurements.

MATERIALS

1   Variable dc power supply
1   VOM
1   DMM
1   10 kΩ resistor
2   100 kΩ resistors

Note: You will need the documentation on your meters to determine their Z_{in} ratings.

1 470 kΩ resistor
1 500 kΩ potentiometer

PROCEDURE

Part I. Meter Input Impedance Ratings

1. Using the documentation on your VOM, determine its input imped-
    ance when it is set on the 10 V, 20 V, or 30 V scale (whichever scale is
    provided by your meter). Record the appropriate scale and the meter's
    corresponding input impedance.
    Scale: _____ Z_{in} (rated): _____
2. Using the documentation on your DMM, determine its input imped-
    ance. Record this value below.
    Z_{in} (rated): _____

Part II. Measuring VOM Input Impedance

3. Construct the circuit shown in Figure 1.2. Set the value of V_S so that it
    is equal to the voltage scale recorded in step 1.
4. Vary the setting of the potentiometer until the meter reads exactly
    one-half the value of V_S. When this occurs, half of V_S is being dropped
    across R_1 and half is being dropped across the input impedance of the
    VOM. This means that the value of R_1 is equal to the value of Z_{in} for the
    VOM.
5. Remove the potentiometer from the circuit and measure its resistance.
    Be careful not to disturb the setting of the potentiometer during the
    process.
    $R_1 = Z_{in}$ of the VOM = _____

Part III. The Effects of Meter Input Impedance
On Voltage Measurements

6. Figure 1.3 shows the schematic of the circuit you will be using in this
    portion of the exercise. As you can see, R_1 will be fixed at 100 kΩ,
    while R_2 will be changed from 10 kΩ to 100 kΩ, and then 470 kΩ.
    Measure and record the values of the resistors you are using in this
    circuit.

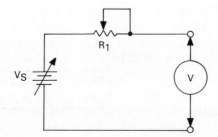

Figure 1.2

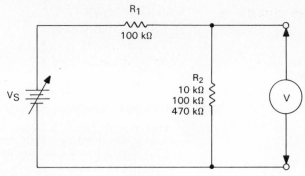

Figure 1.3

Component	Rated value	Measured value
R_1	100 kΩ	_____
R_2	10 kΩ	_____
R_2	100 kΩ	_____
R_2	470 kΩ	_____

> Note: Make sure that you do not confuse R_1 with the 100 kΩ resistor that you will be using as R_2.

7. For each value of R_2 listed in Table 1.1 predict (calculate) the value of V_{R2}. Record the predicted values of V_{R2} in the appropriate spaces in the table.

TABLE 1.1.

R_2	V_{R2} Predicted	Measured—VOM	Measured—DMM
10 kΩ	_____	_____	_____
100 kΩ	_____	_____	_____
470 kΩ	_____	_____	_____

8. Using R_2 = 10 kΩ, measure V_{R2} with the VOM and the DMM. Record the measured values in the appropriate spaces in Table 1.1.

9. Repeat step 8 for R_2 = 100 kΩ and R_2 = 470 kΩ.

THE BRAIN DRAIN (Optional)

10. When you know the value of Z_{in} for a given meter, it is possible to predict the effect that the meter will have on the measured voltage across a given resistance. Using the circuit in Figure 1.1 as a model, determine a method for predicting the loading effect of your VOM. Then, use this method to predict the value of $V_{R2(M)}$ when R_2 = 570 kΩ.

11. Modify the circuit in Figure 1.3 to prove your predicted value of $V_{R2(M)}$ to be within acceptable tolerance limits.

12. In a separate lab report, explain your method for predicting the value of $V_{R2(M)}$. Also, show the circuit you used to measure the value of V_{R2}, and the actual value of $V_{R2(M)}$.

QUESTIONS/PROBLEMS

1. The percent of error between the predicted value of V_{R2} and its measured value is found as

$$\% \text{ of error} = \frac{V_{R2} - V_{R2(M)}}{V_{R2}} \times 100$$

where

$$V_{R2} = \text{the } \textit{predicted} \text{ value of } V_{R2}$$

$$V_{R2(M)} = \text{the } \textit{measured} \text{ value of } V_{R2}$$

Using this equation and the values in Table 1, complete Table 1.2.

TABLE 1.2.

	% of error caused by	
R_2	the VOM	the DMM
10 kΩ	_____	_____
100 kΩ	_____	_____
470 kΩ	_____	_____

2. At which resistance value, if any, did the percent of error caused by your VOM exceed acceptable limits?

3. At which resistance value, if any, did the percent of error caused by your DMM exceed acceptable limits?

4. Your meter input impedance determines the maximum value that R_2 can assume while getting values of $V_{R2(M)}$ that are still within tolerance. For the DMM you used in this exercise, what is the maximum allowable value of R_2? (You'll have to do some creative calculating on this one.)

R_2 (maximum) = _____

5. Discuss, in your own words, what you observed in this exercise.

Exercise 2

Meter Frequency Limitations

OBJECTIVES

- To demonstrate the effects of operating frequency on voltmeter readings.
- To compare the frequency response of a VOM with that of a DMM.

DISCUSSION

You saw in Exercise 1 that the input impedance of a voltmeter can have an impact on the accuracy of your voltage measurements. Another factor that can affect the accuracy of your measurements is the operating frequency of the circuit under test.

AC voltmeters have frequency limits. These limits are usually listed in the documentation of the meter. When a voltmeter is used to measure signals that are outside of its frequency range, any readings obtained will not be accurate. In fact, as the frequency increases beyond the limits of the voltmeter, the readings become less and less accurate.

When a circuit is being operated beyond the frequency limits of your VOM or DMM, you must use an oscilloscope to obtain accurate ac voltage measurements. Also, in most cases, when the circuit under test has *non-sinusoidal* waveforms, such as square waves or triangular waves, you must use an oscilloscope to obtain accurate voltage readings. The exception to this is the *"true rms"* meter, which can accurately measure the rms value of such waveforms.

LAB PREPARATION

Review the material in your basic electronics textbook on the ac voltmeter and its operating frequency limitations.

LAB OVERVIEW

In this exercise, you will:

1. Determine the frequency limits of your VOM and your DMM, using the documentation on the meters.
2. Measure the output voltage of an ac circuit that is being operated at increasing frequencies using both the VOM and the DMM.
3. Graph the *frequency versus voltage reading* characteristics of the VOM and the DMM.

MATERIALS

1 Variable ac signal source
1 VOM
1 DMM
1 100 kΩ resistor

Note: You will need the documentation on your meters to determine their frequency limits.

PROCEDURE

1. Using the documentation on your meters, record their upper frequency limits in the spaces provided.
   VOM frequency limit: _____ (Rated)
   DMM frequency limit: _____ (Rated)
2. Construct the circuit in Figure 2.1.
3. Adjust the amplitude of your signal generator for a DMM reading of 5 Vac. Set the operating frequency to an initial value of 1 kHz. (You do not need to measure the *exact* operating frequency of your signal source. The settings on the frequency control will be considered to be accurate enough for our purposes.)
4. Table 2.1 shows a series of frequency settings. For each frequency listed, measure V_{R1} with both the VOM and the DMM.

Note: Do not change the amplitude setting of your ac signal generator.

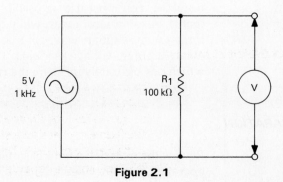

Figure 2.1

Table 2.1

	VOM reading	DMM reading
1 KHz	_____	_____
10 kHz	_____	_____
100 kHz	_____	_____
1 MHz	_____	

5.   Plot the frequency response curves for the VOM and the DMM in the space provided in Figure 2.2.

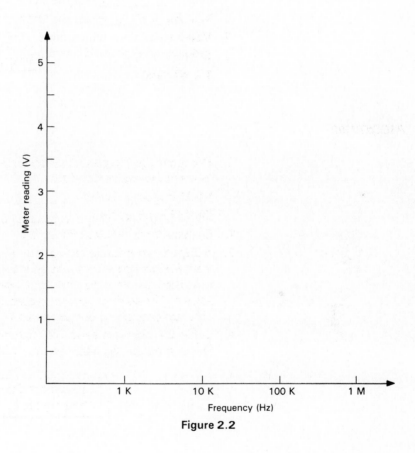

Figure 2.2

THE BRAIN DRAIN (Optional)

6.   In this exercise, the circuit resistance was set to 100 kΩ. Devise a method for testing the frequency response of each of your meters when R_1 equals the input impedance of the meter under test. Then, measure the frequency response of each of your meters under this condition and compare the results to those obtained in steps 4 and 5 of the procedure.
7.   Write a separate report showing your procedure and results for step 6.

QUESTIONS/PROBLEMS

1. Using your graph from step 5 in the procedure, determine the value of V_R when your circuit is operating at the frequency limits for the VOM and the DMM (see the value recorded in step 1 of the procedure).

   $V_R = $ _____ when the VOM is operated at its frequency limit

   $V_R = $ _____ when the DMM is operated at its frequency limit

2. What percent of error is introduced by operating each of the meters at its frequency limit?

   % of error = _____ at the VOM frequency limit

   % of error = _____ at the DMM frequency limit

3. If we consider an error of ten percent to be acceptable, what is the frequency limit of each of the meters?

   $f_{limit} = $ _____ for the VOM

   $f_{limit} = $ _____ for the DMM

4. In your opinion, which meter (the VOM or the DMM) has the better frequency response characteristics? Explain your answer.

5. Discuss, in your own words, what you observed in this exercise.

PART II

Diodes and Diode Circuits

Exercise 6

Full-Wave Center-Tapped Rectifiers

OBJECTIVES

- To demonstrate the operation of the full-wave center-tapped rectifier.
- To provide the opportunity for the practical analysis of a simple full-wave center-tapped rectifier.
- To demonstrate some of the common fault symptoms that may occur in a full-wave center tapped rectifier.

DISCUSSION

The full-wave center-tapped rectifier changes ac to pulsating dc by either converting the negative alternations to positive alternations (for a positive dc power supply) or by converting the positive alternations of the input signal to negative alternations (for a negative dc power supply). The input and output waveforms for a *positive* full-wave center-tapped rectifier are shown in Figure 6.1.

LAB PREPARATION

Review section 3.2 of *Introductory Electronic Devices and Circuits*.

LAB OVERVIEW

In this exercise, you will:

1. Predict (calculate) the peak and average output values for a *positive* full-wave center-tapped rectifier using the measured rms secondary voltage of the transformer.

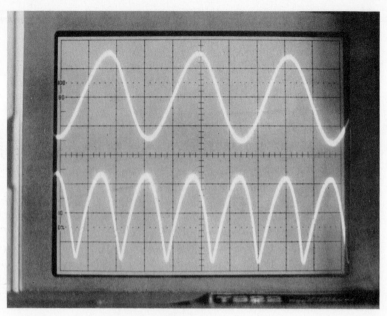

Figure 6.1

2. Observe the phase relationship between the two end leads of the transformer secondary.
3. Measure the peak and average peak and average voltages for the full-wave center-tapped rectifier.
4. Predict and measure the peak and average voltage values for a *negative* full-wave center-tapped rectifier.
5. Observe some of the common fault symptoms that occur in the full-wave center-tapped rectifier.

MATERIALS

1 VOM or DMM
1 Dual-trace oscilloscope
1 Center-tapped transformer (rated between 12 Vac and 24 Vac)
2 1N4001 rectifier diodes
1 5.6 kΩ resistor

PROCEDURE

1. Construct the circuit shown in Figure 6.2.
2. Apply power to the circuit and measure the rms transformer secondary voltage.

   V_2 = _____ Vac

3. Using the value obtained in step 2, predict the peak load voltage for the circuit.

   $V_{out(pk)}$ = _____ V

4. Using the value calculated in step 3, predict the *average* (dc) load voltage for the circuit.

   V_{ave} = _____ Vdc

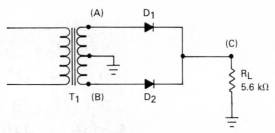

Figure 6.2

5. Establish the center of your oscilloscope graticule as being the ground position for both traces, as follows:
    a. Set both AC/GND/DC switches in the GND position.
    b. Using the vertical position controls, adjust both traces until they are on the center line of the graticule.
    c. Return both AC/GND/DC switches to the AC position.
6. Connect channel A of your oscilloscope to point (A) in the circuit and connect channel B to point (B) in the circuit. Draw the waveforms you see on the CRT as neatly as possible in the space provided below.

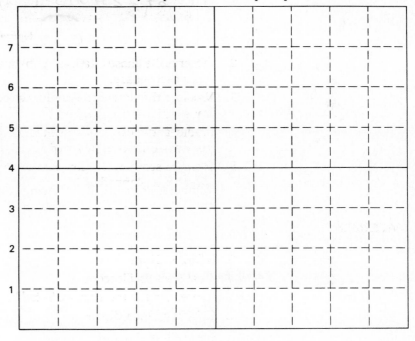

Time/Div:_____   V/Div:_____

7. Based on the waveforms you are seeing, what conclusion can you draw about the phase relationship of the waveforms at points (A) and (B)?
8. Set your oscilloscope for DC coupling and view the waveform at point (C) in the circuit. Draw this waveform as neatly as possible in the space provided on the next page.
9. Using a dc voltmeter, measure the average (dc) load voltage.
    V_{ave} = _____ Vdc
10. Disconnect power from the circuit and reverse the direction of *both* diodes. Draw the schematic diagram for the new circuit in the space provided below.

**Step 8
Waveform**

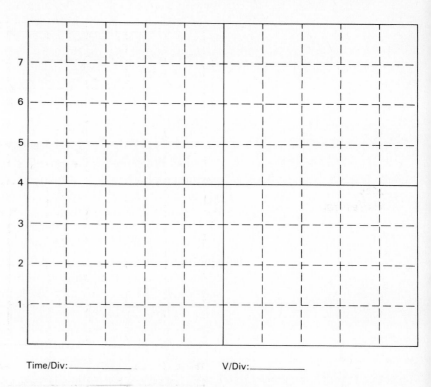

Time/Div:_____   V/Div:_____

11. Predict the following values for the circuit you have drawn.

$V_{out(pk)}$ = _____

V_{ave} = _____

Note: Use the measured transformer rms secondary voltage in
your calculations.

12. Using the oscilloscope and the
voltmeter, measure the following voltages:

$V_{out(pk)}$ = _____

V_{ave} = _____

Part II. Fault Symptoms

13. Disconnect power from the circuit. Then, remove D_1 from the circuit.
This simulates an *open diode* in the rectifier. After removing D_1, reapply
power to the circuit.

Note: Whenever you are directed to remove
a component, a gap should be left where the
component appeared in the circuit. Do not bridge
the gap left by the missing component unless
directed to do so.

14. Observe the waveform at the circuit output. Draw this waveform in the
space provided.

15. Disconnect power from the circuit, return D_1 to its original position,
and remove the load resistor. Then, reapply power to the circuit.

16. Observe the waveform at point (C) in the circuit. Draw this waveform
in the space provided.

**Step 14
Waveform**

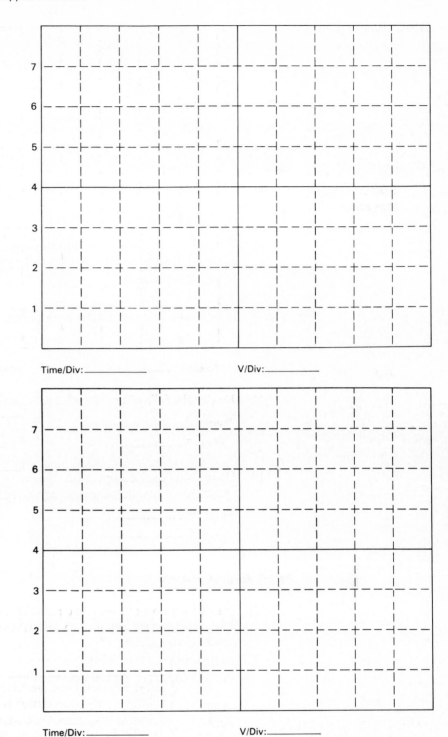

Time/Div:_____   V/Div:_____

**Step 16
Waveform**

Time/Div:_____   V/Div:_____

THE BRAIN DRAIN (Optional)

17. Predict and measure the values of PIV and $P_{D(max)}$ for the diodes in your rectifier. In a separate report, show the method you used to measure these values, along with the predicted and measured voltage and power dissipation.

QUESTIONS/PROBLEMS

1. What happens to the full-wave rectifier when you reverse the diode positions?

2. For each set of voltage readings in steps 11 and 12, calculate the percent of error between your calculated and measured values.

   $V_{out(pk)}$: % of error = _____

   V_{ave}: % of error = _____

3. How would you explain the error percentages in question 2?

4. Very often, when one diode opens in a full-wave rectifier, the other diode is opened in the process. If both diodes in Figure 6.2, what kind of an output (if any) would you expect to see from the rectifier? Explain your answer.

5. Discuss, in your own words, what you observed in this exercise. (Use a separate sheet of paper, if necessary.)

Exercise 8

Rectifiers: A Comparison

OBJECTIVES

- To provide the opportunity to observe the similarities and differences between the three rectifier circuits.
- To provide the opportunity to review the last three exercises on rectifier circuits.

DISCUSSION

In this exercise, you will not be taking any circuit measurements. Rather, you will be comparing the results that you have obtained in exercises 5 through 7. By comparing the results of these three exercises, you will be able to better understand the differences and similarities that exist between the three commonly used rectifier circuits.

LAB PREPARATION

Make sure that you have completed all of the waveform drawings, calculations, and questions for exercises 5 through 7.

LAB OVERVIEW

In this exercise, you will:

1. Compare and contrast the output waveforms for the half-wave, center-tapped, and bridge rectifiers you analyzed in exercises 5 through 7.
2. Compare and contrast the measured peak and average output voltage values for the rectifiers you analyzed in exercises 5 through 7.

PROCEDURE

1. Refer back to exercises 5 through 7. Locate the normal output waveforms for each of the rectifier circuits and neatly draw these waveforms in the spaces provided.

> Note: Use the *positive* half-wave rectifier and *positive* full-wave center-tapped rectifier output waveforms.

Half-Wave Rectifier:

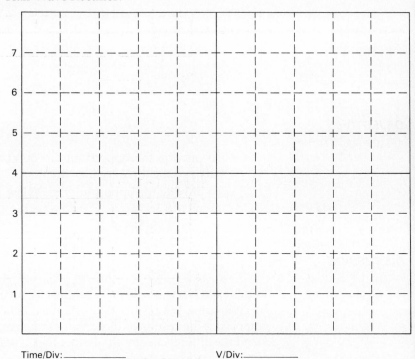

Time/Div: _____   V/Div: _____

2. Complete Table 8.1 using the *measured* values in exercises 5 through 7.

TABLE 8.1

Value	Half-wave	Full-wave C.T.	Bridge
		Rectifier type:	
$V_{out(pk)}$	_____	_____	_____
V_{ave}	_____	_____	_____

Full-Wave Center-Tapped Rectifier:

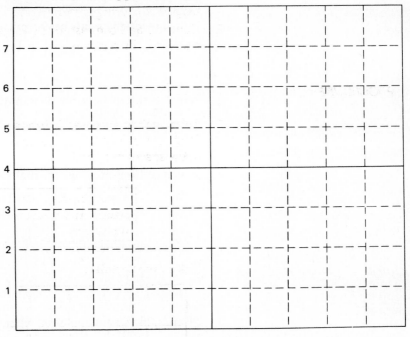

Time/Div:_____   V/Div:_____

Bridge Rectifier:

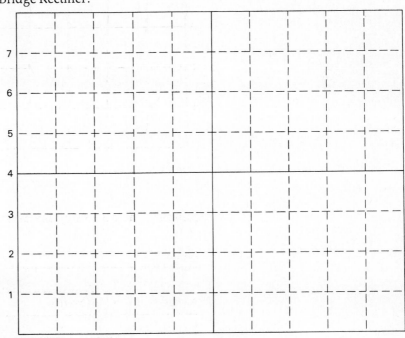

Time/Div:_____   V/Div:_____

QUESTIONS/PROBLEMS

1. Which rectifier circuits have similar output waveforms?

2. Which rectifiers have nearly identical peak output voltages?

3. Which rectifier provides the highest dc output voltage?

4. Which rectifier provides the lowest dc output voltage?

5. Discuss, in your own words, what you observed in this exercise.

$$\frac{5.0 - 4.3 V}{5.0 V}$$

Exercise 13

Voltage Multipliers

OBJECTIVES

- To demonstrate the operation of the half-wave and full-wave voltage doublers.
- To demonstrate the operation of voltage triplers and voltage quadruplers.
- To demonstrate the use of the voltage multiplier in creating a dual-polarity dc power supply.

DISCUSSION

A voltage multiplier is a diode circuit that is used to provide a dc output that is a specified multiple of the peak value of its ac input voltage. For example, the dc output from a voltage *doubler* is approximately two times its peak input voltage. The voltage *tripler* has a dc output that is approximately three times its peak input voltage, and so on.

Among other applications, a voltage quadrupler can be used to create a dual-polarity power supply. That is, a dc power supply that provides both positive and negative dc output voltages. This application of the voltage quadrupler is demonstrated in this exercise.

LAB PREPARATION

Review section 4.4 of *Introductory Electronic Devices and Circuits.*

LAB OVERVIEW

In this exercise, you will:

1. Construct a half-wave voltage doubler and observe its input/output characteristics.
2. Construct a full-wave voltage doubler and observe its input/output characteristics.
3. Construct a voltage tripler and observe its input/output characteristics.
4. Construct a voltage quadrupler and observe its input/output characteristics.
5. Construct and analyze a dual-polarity dc power supply.

MATERIALS

1   Dual-trace oscilloscope
1   VOM or DMM
1   12 Vac transformer

> Note: If you have a 24 Vac center-tapped transformer, you can connect the circuits between one of the transformer end leads and the center tap.

4   1N4001 rectifier diodes
1   10 Ω resistor
4   47 μF electrolytic capacitors, 100 Vdc rated
1   100 μF electrolytic capacitor, 100 Vdc rated
1   10 kΩ potentiometer (optional)

PROCEDURE

1. Construct the half-wave voltage doubler shown in Figure 13.1.
2. Using your oscilloscope, measure and record the peak secondary voltage of the transformer.
   $V_{2(pk)}$ = _____
3. Using your voltmeter, measure the dc output voltage from the circuit.
   Vdc = _____

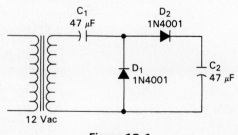

Figure 13.1

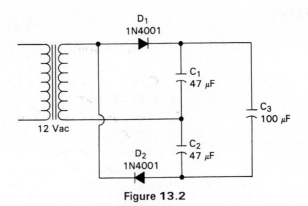

Figure 13.2

4. Set your oscilloscope for ac coupling. Measure and record the peak-to-peak ripple output voltage from the circuit.

$V_{r(out)} = $ _____ V_{PP}

5. Construct the full-wave voltage doubler shown in Figure 13.2.

> Note: Initially, C_3 should be left out of
> the circuit. It will be added in step 8.

6. Using your voltmeter, measure and record the dc output voltage from the circuit.

$Vdc = $ _____

How does this value compare with the one obtained in step 3?

7. Set your oscilloscope for ac coupling. Measure and record the peak-to-peak ripple output voltage from the circuit.

$V_{r(out)} = $ _____ V_{PP}

How does this value compare with the one obtained in step 4?

8. Disconnect the input signal from the circuit and insert C_3 at the point indicated.

> Caution: The capacitors in the
> voltage multipliers are being
> used to store a charge. Before
> removing a given capacitor
> from the circuit, connect your
> 10 Ω resistor across the
> capacitor leads to bleed off the charge.

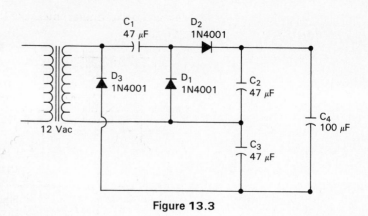

Figure 13.3

9. Reapply the circuit input signal. Measure and record the peak-to-peak ripple output voltage from the circuit.

$V_{r(out)} =$ _____ V_{PP}

How does this value compare with the one obtained in step 7?

10. Construct the voltage tripler shown in Figure 13.3.
11. Using your voltmeter, measure the dc output voltage from the circuit.

    $Vdc =$ _____

12. Set your oscilloscope for ac coupling. Measure and record the peak-to-peak ripple output voltage from the circuit.

    $V_{r(out)} =$ _____ V_{PP}

13. Construct the voltage quadrupler shown in Figure 13.4.
14. Using your voltmeter, measure the dc output voltage from the circuit.

    $Vdc =$ _____

15. Set your oscilloscope for ac coupling. Measure and record the peak-to-peak ripple output voltage from the circuit.

    $V_{r(out)} =$ _____ V_{PP}

16. Modify the circuit in Figure 13.4 as shown in Figure 13.5.
17. Using your voltmeter, measure the dc output voltage from point (A) to ground.

    $Vdc =$ _____

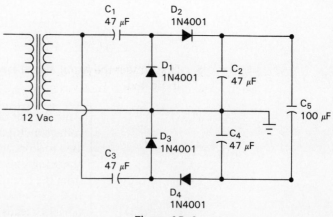

Figure 13.4

18. Using your voltmeter, measure the dc output voltage from point (B) to ground.

    Vdc = _____

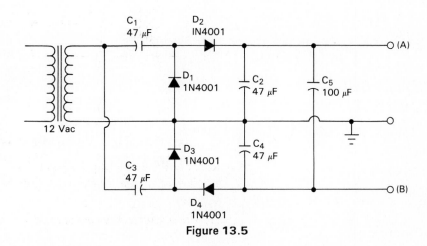

C_1
47 μF

D_2
IN4001

D_1
1N4001

C_2
47 μF

C_5
100 μF

(A)

12 Vac

D_3
1N4001

C_4
47 μF

C_3
47 μF

D_4
1N4001

(B)

Figure 13.5

THE BRAIN DRAIN (Optional)

19. Throughout this exercise, we have ignored the effects that a load resistance can have on the output ripple voltage from a voltage multiplier. Using a 10 kΩ potentiometer, analyze the effects of load resistance on the amplitude of the output ripple from a full-wave voltage doubler.

20. In a separate report, include the values of load resistance used and the corresponding values of $V_{r(out)}$. Also, plot a curve showing load resistance versus $V_{r(out)}$ with C being held constant.

QUESTIONS/PROBLEMS

1. Compare the value of Vdc in step 3 with the value of $V_{2(pk)}$ in step 2. How do these values compare?

2. Compare the value of Vdc in step 6 of the procedure with the value of $V_{2(pk)}$ in step 2. How do these values compare?

3. Compare the value of Vdc in step 11 of the procedure with the value of $V_{2(pk)}$ in step 2. How do these values compare?

4. Compare the value of Vdc in step 14 of the procedure with the value of $V_{2(pk)}$ in step 2. How do these values compare?

5. Step 7 of the procedure tells you to compare the ripple output from the full-wave doubler with that of the half-wave doubler. Based on your answer to that question, is there any advantage in using the full-wave doubler over the half-wave doubler?

6. Step 9 of the procedure tells you to compare the ripple output from a filtered multiplier with that of an equivalent multiplier without a filter capacitor. Based on your answer to that question, is there any advantage to using a filter capacitor at the output of a multiplier?

7. Discuss, in your own words, what you observed in this exercise.

PART III

BJTs And BJT Circuits

Exercise 14

BJT Current and Voltage Characteristics

- To demonstrate the relationship between collector current (I_C) and base current (I_B).
- To provide the opportunity for plotting the output characteristic curves for a transistor using measured component values.

DISCUSSION

The BJT is a three-terminal device whose collector current (I_C) is controlled by the base current (I_B). As I_B is changed, I_C changes proportionally, as given in the relationship

$$I_C = h_{FE}I_B$$

where h_{FE} is the dc base-to- collector current gain of the component.

The voltage across the collector and emitter terminals of the BJT (V_{CE}) is determined by the collector supply voltage (V_{CC}), the total resistance in the collector and emitter circuits (R_C and R_E), and the value of I_C.

LAB PREPARATION

Review Chapter 5 of *Introductory Electronic Devices and Circuits*.

LAB OVERVIEW

In this exercise, you will:

1. Adjust the base resistance (R_B) of a BJT circuit to provide a specified value of I_B.
2. Adjust V_{CC} to provide a series of V_{CE} values, and measure the value of I_C at each V_{CE} interval.
3. Repeat steps 1 and 2 for a series of I_B values.
4. Using the collected data, plot the collector curves for the BJT under test.

MATERIALS

2 Variable dc power supplies
3 VOMs and/or DMMs
1 2N3904 npn transistor
1 100 kΩ resistor

> Note: At least one of the three meters should be a DMM.

PROCEDURE

1. Measure and record the actual resistance of your 100 kΩ resistor.
   $R_1 =$ _____
2. Construct the circuit shown in Figure 14.1. Both supply voltages should initially be set to 0 Vdc.
3. Calculate the value of V_{BB} that is needed to have 5 μA of current through R_1.
   $V_1 =$ _____ when $I_B = 5$ μA.
4. Adjust V_{BB} to obtain the value of V_1 calculated in step 3.

> Note: You should use a DMM to measure V_1 because of the relatively high value of R_1.

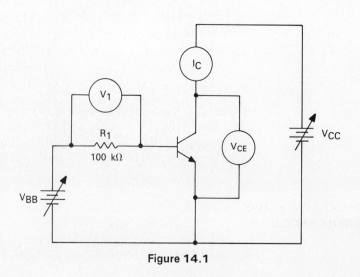

Figure 14.1

5. Adjust V_{CC} so that V_{CE} is 0.5 Vdc.

6. Measure and record the value of I_C in the appropriate space in Table 14.1.

7. Increase V_{CC} to provide a V_{CE} of 1 Vdc. Measure and record the corresponding value of I_C in Table 14.1.

8. Repeat step 7 for each of the values of V_{CE} listed in Table 14.1.

TABLE 14.1

I_B (μA)	V_{CE} (Vdc)					
	+ 0.5	+ 1	+ 5	+ 10	+ 15	+ 20
5	——	——	——	——	——	——
10	——	——	——	——	——	——
20	——	——	——	——	——	——
30	——	——	——	——	——	——
40	——	——	——	——	——	——
50	——	——	——	——	——	——

9. After completing the measurements in step 8 for $I_B = 5$ μA, return V_{CE} to 0 Vdc. Then, adjust V_{BB} to provide the value of V_1 needed to cause I_B to equal 10 μA.

10. Repeat steps 5 through 8 for $I_B = 10$ μA.

11. Repeat steps 5 through 10 until Table 14.1 is complete.

12. Using the data recorded in Table 14.1, plot the characteristic collector curves for the 2N3904 in the space provided in Figure 14.2. Your curves should be similar to those shown in Figure 5.27 in *Introductory Electronic Devices and Circuits*.

THE BRAIN DRAIN (OPTIONAL)

13. Figure 5.28 in *Introductory Electronic Devices and Circuits* shows the typical base curve for a BJT. Devise a procedure for obtaining a series of I_B versus V_{BE} combinations, and use that procedure to obtain the values needed to plot the actual base curve for your 2N3904. Also, prove that the base curve is relatively independent of the value of V_{CE} for the transistor.

14. In a separate report, show your procedure for obtaining the I_B versus V_{BE} values, the actual values obtained, any relevant circuit schematics, and the plotted base curve for the 2N3904. Also, include your analysis of the relationship between the 2N3904 base curve and the value of V_{CE} for the device.

QUESTIONS/PROBLEMS

1. What happened to the value of I_C as you increased V_{CE} for a given value of I_B?

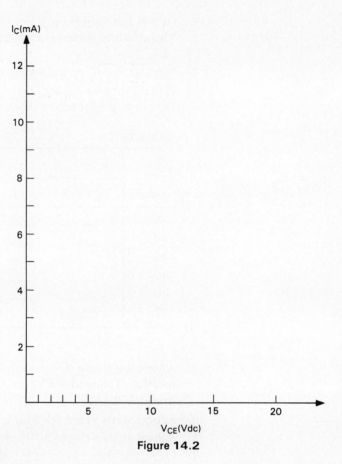

Figure 14.2

2. What happened to I_C as you increased the value of I_B?

3. The dc current gain (h_{FE}) is equal to the ratio of I_C to I_B. Determine the value of h_{FE} for your transistor at the point on your collector characteristic curves where $V_{CE} = 10$ Vdc and $I_B = 30$ μA.

    $h_{FE} =$ _____

4. Discuss the similarities and differences between your transistor characteristic collector curves and those shown in Figure 5.27 of *Introductory Electronic Devices and Circuits*.

5. Using *your* characteristic curves, predict the values of I_C for each of the I_B and V_{CE} combinations listed below.

   $I_C =$ _____ when $I_B = 25\ \mu A$ and $V_{CE} = 10$ Vdc.

   $I_C =$ _____ when $I_B = 35\ \mu A$ and $V_{CE} = 8$ Vdc.

6. Discuss, in your own words, what you observed in this exercise.

Exercise 15

Base Bias

- To demonstrate the dc operation of the base bias circuit.
- To provide the opportunity for a load line analysis of a base bias circuit.
- To demonstrate the effect of a change in base current on the Q point of a transistor dc biasing circuit.
- To demonstrate the relative instability of the base bias circuit.

DISCUSSION

Base bias is the simplest of the transistor biasing circuits. It consists of a single transistor, two resistors, and a single dc power supply.

It would seem that the simplicity of the base bias circuit would make it ideal for most applications. However, the base bias circuit is relatively unstable. That is, the Q-point of the circuit will *shift* (change) if there is a significant change in h_{FE} and/or temperature. This point will be demonstrated in this exercise.

LAB PREPARATION

Review sections 6.1 and 6.2 of *Introductory Electronic Devices and Circuits*.

LAB OVERVIEW

In this exercise, you will:

1. Plot the dc load line for a base bias circuit using its measured resistor values, and use this load line to predict a series of I_C and V_{CE} combinations for the circuit.

2. Verify the I_C versus V_{CE} combinations predicted for the circuit.

3. Change the BJT being used in the circuit to observe the effects of a change in h_{FE} on the Q-point of the base bias circuit.

4. Cause a change in temperature and observe its effects on the Q-point of the base bias circuit.

MATERIALS

1   Variable dc power supply
2   VOMs and/or DMMs
2   2N3904 npn transistors
2   Resistors: 2 kΩ and 10 kΩ
1   2.5 MΩ potentiometer
1   Soldering iron
2   2N3906 pnp transistors (optional)

PROCEDURE

Part I. The Base Bias Circuit

1. Measure and record the actual values of your fixed resistors.
    R_C (2 kΩ rated)   = _____
    R_{B2} (10 kΩ rated) = _____

2. Construct the circuit shown in Figure 15.1

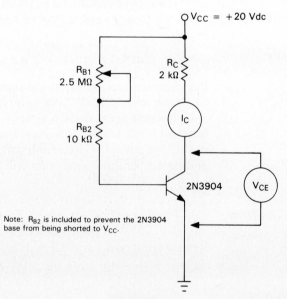

Note: R_{B2} is included to prevent the 2N3904 base from being shorted to V_{CC}.

Figure 15.1

3. Using your measured value of R_C and V_{CC} = 20 Vdc, plot the load line for the circuit in the space provided in Figure 15.2.

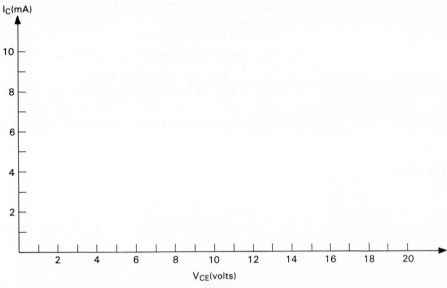

Figure 15.2

4. Using your plotted load line, determine the approximate values of V_{CE} and I_C for the circuit when it is operated at midpoint.

V_{CE} = _____ for the midpoint bias
I_C = _____ for the midpoint bias

5. Set R_{B1} to its maximum value and apply power to the circuit.

6. Adjust R_{B1} until V_{CE} is equal to the value obtained in step 4. Measure and record the value of I_C at this setting.

I_C = _____

How does this value of I_C compare with the value predicted in step 4?

7. Using your plotted load line, predict the value of I_C when V_{CE} = 2 Vdc.

I_C = _____

8. Adjust R_{B1} until V_{CE} is equal to 2 Vdc. Measure and record the corresponding value of I_C.

I_C = _____

How does this value of I_C compare with the value predicted in step 7?

9. Using your plotted load line, predict the value of I_C when V_{CE} = 16 Vdc.

I_C = _____

10. Adjust R_{B1} until V_{CE} is equal to 16 Vdc. Measure and record the corresponding value of I_C.

I_C = _____

How does this value of I_C compare with the value predicted in step 9?

Part II. Base Bias Instability

11. Adjust R_{B1} until the circuit is once again at midpoint bias.
12. Disconnect power from the circuit and swap 2N3904 transistors.

> Note: Do not adjust R_{B1} until directed to do so.

13. Reapply power to the circuit and observe the effects that changing the transistor had on your Q point values of V_{CE} and I_C. Note the changes in the space provided.

How would you account for the changes in V_{CE} and I_C? In other words, why would a change in 2N3904s have the effects that you observed?

14. Readjust R_{B1} to once again obtain midpoint bias.
15. Heat up your soldering iron. While it is heating, keep an eye on V_{CE}. If it wanders from its midpoint value, adjust R_{B1} to compensate for the change.
16. When your soldering iron gets hot, *carefully* touch it to the emitter terminal of your transistor and observe what happens to V_{CE}.

> Note: Touch the emitter lead with the soldering iron only for as long as needed to see a change in V_{CE}!

Record your observations below.

THE BRAIN DRAIN (Optional)

17. The 2N3906 is the pnp equivalent of the 2N3904. Perform the complete load line analysis (including the Q-point shifts caused by changes in h_{FE} and temperature) for the 2N3906.
18. In a separate report, show your procedure for analyzing the load line characteristics of the 2N3906, all relevant schematics and diagrams, and a conclusion on the similarities between the 2N3906 and the 2N3904.

QUESTIONS/PROBLEMS

1. I_C can be calculated using the equation:

$$I_C = \frac{V_{CC} - V_{CE}}{R_C}$$

Using this equation and your measured value of R_C, calculate the following:

$I_C =$ _____ when $V_{CE} = 2$ Vdc

$I_C =$ _____ when $V_{CE} = 16$ Vdc

How do these values of I_C compare with those measured in steps 8 and 10 of the procedure?

2. V_{CE} *decreased* when you heated your transistor. Why?

3. In Figure 15.2, plot the points that you measured in steps 6, 8, and 10 of the procedure.
   Discuss why your points do or do not fall on the load line.

4. Based on your observations in this exercise, what is your opinion of base bias as a means of providing a stable Q point?

5. Discuss, in your own words, what you observed in this exercise.

Exercise 16

Emitter Bias

OBJECTIVES

- To demonstrate the dc operating characteristics of emitter bias.
- To demonstrate the relative Q-point stability of the emitter bias circuit.
- To demonstrate the typical fault symptoms that may occur in an emitter bias circuit.

DISCUSSION

The emitter bias circuit uses a dual-polarity power supply to bias its BJT. Under normal circumstances, the transistor emitter terminal is at some voltage that is slightly negative, usually between -0.7 and -1 V. The base terminal of the transistor is usually at or near ground, and the collector terminal is at a voltage that is approximately equal to one-half of V_{CC} when the circuit is midpoint biased.

Since it requires a dual-polarity power supply, the emitter bias circuit is not used as often as many of the other common dc biasing circuits. However, it *is* used often enough to merit a look at its operation and fault symptoms.

LAB PREPARATION

Review section 6.3 of *Introductory Electronic Devices and Circuits*.

LAB OVERVIEW

In this exercise, you will:

1. Construct an emitter bias circuit and adjust it for midpoint bias.
2. Measure the values of V_{CEQ}, V_C, V_B, V_E, and I_{CQ} for the midpoint biased amplifier.
3. Test the Q-point stability of the circuit.
4. Introduce a series of faults into the circuit and observe the effects of these faults on the values of V_C, V_B, and V_E.

MATERIALS

1 Dual-polarity variable dc power supply
1 VOM
1 DMM
2 2N3904 npn transistors
3 Resistors: 6.8 kΩ, 10 kΩ, and 11 kΩ
1 50 kΩ potentiometer
1 Soldering iron

Note: You may use a second DMM in place of the VOM, if desired.

PROCEDURE

1. Construct the circuit shown in Figure 16.1.
2. Apply power to the circuit and adjust R_{B1} so that the circuit is midpoint biased; that is, so that V_{CEQ} is approximately one-half of V_{CC}. Record your measured value of V_{CEQ}.

   $V_{CEQ} = $ _____
3. Measure and record the Q-point value of I_C.

   $I_{CQ} = $ _____
4. Measure and record the following voltages:

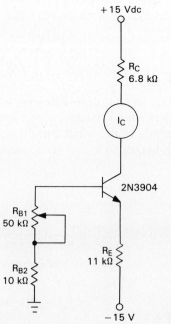

Figure 16.1

$V_C =$ _____

$V_E =$ _____

$V_B =$ _____

> Note: Be sure that you measure these voltages *with respect to ground*.

Part II. Bias Stability

5. Check to make sure that your circuit is still midpoint biased. If it isn't, adjust R_{B1} to return the circuit to midpoint operation.

6. Disconnect power from the circuit and swap 2N3904 transistors.

> Note: Do not adjust R_{B1} again until directed to do so.

7. Reapply power to the circuit and observe the effects, if any, that changing the transistor had on your values of V_{CEQ} and I_{CQ}. Note the changes (or lack of changes) in the space provided.

Why did (or didn't) changing your transistor cause the values of V_{CEQ} and I_{CQ} to change?

8. If necessary, adjust R_{B1} to again provide midpoint bias.

9. Heat up your soldering iron. While it is heating, keep an eye on V_{CEQ} and I_{CQ}. If they wander from their midpoint values, adjust R_{B1} to compensate for the changes.

10. When your soldering iron gets hot, *carefully* touch it to the emitter lead of your transistor and observe what, if anything, happens to V_{CEQ} and I_{CQ}.

> Note: Do not touch the soldering iron to the emitter lead for more than a couple of seconds!

Record your observations below.

Part III. Fault Symptoms

11. Allow the circuit to cool and adjust R_{B1} to again provide midpoint bias, if necessary.

12. With the power applied, remove R_C from the circuit.

> Note: Whenever you are directed to remove a component, a gap should be left where the component appeared in the circuit. Do not bridge the gap left by the component unless directed to do so.

Measure and record the following values:

V_C = _____

V_E = _____

V_B = _____

I_C = _____

13. Return R_C to its original position in the circuit and remove R_E. Measure and record the following values:

V_C = _____

V_E = _____

V_B = _____

I_C = _____

14. Return R_E to its original position in the circuit and remove R_{B2}. Measure and record the following values:

V_C = _____

V_E = _____

V_B = _____

I_C = _____

THE BRAIN DRAIN (Optional)

15. Plot the dc load line for the circuit in Figure 16.1. Then, using at least 5 sets of circuit measurements, verify the validity of the load line.

16. In a separate report, show your load line, your predicted and measured V_{CEQ} and I_{CQ} combinations, and an explanation of any points that fail to fall on the load line.

QUESTIONS/PROBLEMS

1. If you performed Exercise 15, compare your observations in procedure step 13 of that exercise with those in procedure step 7 of this exercise. Which circuit had the largest changes in V_{CEQ} and I_{CQ} when your transistors were changed?

How would you account for the difference between the stability of base bias and that of emitter bias?

2. If you performed Exercise 15, compare your observations in procedure step 16 of that exercise with those in procedure step 10 of this exercise. Which circuit had the largest changes in V_{CEQ} and I_{CQ} when temperature was increased?

3. Explain the fault symptoms you observed in step 12.

4. Explain the fault symptoms you observed in step 13.

5. Explain the fault symptoms you observed in step 14.

6. Discuss, in your own words, what you observed in this exercise.

$$V_B = \frac{R_1 + R_2}{R_2} \, V_{CC}$$

$$\frac{R_2}{R_1 + R_2}$$

Exercise 18

Bias Stability

OBJECTIVE

- To further demonstrate the concept of bias stability.

DISCUSSION

The circuit shown in Figure 18.1a has good bias stability. Good bias stability is achieved by designing the circuit so that R_2 is less than one-tenth of the product $h_{FE}R_E$. This simple criteria stabilizes the circuit against changes in h_{FE}.

Care must be taken, however, to protect against variations in V_{BE}. Keeping V_B equal to or greater than $5V_{BE}$ should stabilize the circuit against changes in V_{BE}. The Q-point can also shift due to changes in V_{CC}, since the equivalent base supply voltage, V_{BB}, is obtained by a voltage divider from V_{CC}.

The circuit shown in Figure 18.1b has worse bias stability than the one in Figure 18.1a. This is due to the fact that R_4 is almost equal to $h_{FE}R_E$. Thus, the current through R_4 will be approximately equal to I_B. In this circuit, any change in temperature and/or h_{FE} will cause V_B to change significantly, and thus, will cause a significant shift in the circuit Q-point.

LAB PREPARATION

Review section 6.4 of *Introductory Electronic Devices and Circuits*.

LAB OVERVIEW

In this exercise, you will:

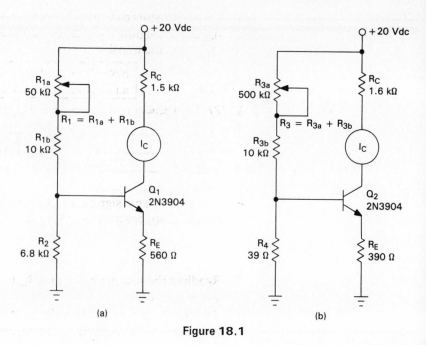

Figure 18.1

1. Construct the two circuits in Figure 18.1 and adjust them both for midpoint bias.

2. Swap transistors between the two circuits and note the effects of the swap on the values of V_{CEQ} and I_{CQ} for each circuit.

3. Replace the transistor in each circuit with a "high h_{FE}" transistor and note the effects of the change on the values of V_{CEQ} and I_{CQ} for each circuit.

MATERIALS

1 Variable dc power supply
3 VOMs and/or DMMs
2 2N3904 npn transistors
1 2N2222 npn transistor
8 Resistors: 390 Ω, 560 Ω, 1.5 kΩ, 1.6 kΩ, 6.8 kΩ, 10 kΩ (2), and 39 kΩ
2 Potentiometers: 50 kΩ and 500 kΩ

> Note: Any combination of VOMs and/or DMMs is acceptable.

PROCEDURE

1. Construct both of the circuits shown in Figure 18.1. Before applying power, set R_1 to 30 kΩ and R_3 to 180 kΩ.

2. Apply power to the circuits. Measure and record the following values for the circuits:

   V_{CEQ1} = _____

   I_{CQ1} = _____

   V_{CEQ2} = _____

   I_{CQ2} = _____

3. Adjust R_{1a} and R_{3a} so that both of the circuits are midpoint biased.

4. Two transistors can have the same part number and still have values of h_{FE} that are significantly different. Remove power from the two circuits and swap the transistors.

> Note: Do not adjust the settings of R_{1a} and R_{3a} until directed to do so.

5. Reapply power to the circuits and measure the following values:

V_{CEQ1} = _____

I_{CQ1} = _____

V_{CEQ2} = _____

I_{CQ2} = _____

6. Were there any significant changes in the Q-point values of either circuit? If so, describe the changes.

7. Readjust the settings of R_{1a} and R_{3a} to return both circuits to midpoint bias if necessary.

8. Replace Q_1 (Figure 18.1a) with the 2N2222 transistor. Note any significant changes in V_{CEQ} and I_{CQ} for the circuit in the space provided.

9. Replace Q_2 (Figure 18.1b) with the 2N2222 transistor. Note any significant changes in V_{CEQ} and I_{CQ} for the circuit in the space provided.

QUESTIONS/PROBLEMS

1. Based on your experience with this exercise, which circuit in Figure 18.1 has the best bias stability?

2. A good estimate of bias stability is the ratio of the value of R_2 to the value of R_E, where a low ratio is an indicator of good bias stability. Calculate these ratios for both circuits in Figure 18.1.

   For Figure 18.1a: $\dfrac{R_2}{R_E}$ = _____

   For Figure 18.1b: $\dfrac{R_4}{R_E}$ = _____

3. Do the results obtained in question 2 agree with your opinion in question 1?

4. Discuss, in your own words, what you observed in this exercise.

Collector-Feedback Bias

OBJECTIVES

- To demonstrate the dc operating characteristics of collector-feedback bias.
- To demonstrate the common fault symptoms that may occur in a collector-feedback biasing circuit.

DISCUSSION

The collector-feedback bias circuit is a relatively stable BJT biasing circuit that utilizes only two resistors and a single dc power supply. The circuit operates to stabilize the Q-point against changes in V_{BE} and h_{FE}. The circuit stability is accomplished by feeding a portion of the dc output current back to the base circuit.

Though the collector-feedback bias circuit is somewhat stable against changes in V_{BE} and h_{FE}, it does not have the stability of the voltage-divider bias circuit. For this reason, it is used less frequently than voltage-divider bias.

LAB PREPARATION

Review section 6.5 of *Introductory Electronic Devices and Circuits*.

LAB OVERVIEW

In this exercise, you will:

1. Construct a collector-feedback bias circuit and observe its operating and stability characteristics.

2. Introduce a series of faults into the circuit and observe the effects that these faults have on the values of V_{CEQ} and I_{CQ}.

MATERIALS

1 Variable dc power supply
3 VOMs and/or DMMs
2 2N3904 npn transistors
2 Resistors: 2.2 kΩ and 10 kΩ
1 500 kΩ potentiometer

Note: Any combination of VOMs and/or DMMs is acceptable.

PROCEDURE

1. Measure and record the values of your fixed resistors.
   R_C (2.2 kΩ rated) = _____
   R_{B2} (10 kΩ rated) = _____
2. Construct the circuit shown in Figure 19.1. Start with R_{B1} adjusted to its maximum value.
3. Apply power to the circuit and adjust R_{B1} to obtain midpoint bias. Record your measured values of V_{CEQ}, I_{CQ}, and I_B.
   V_{CEQ} = _____
   I_{CQ} = _____
   I_B = _____
4. Disconnect power from the circuit and replace the transistor with your other 2N3904.

Note: Do not adjust the setting of R_{B1} until directed to do so.

5. Reapply power to the circuit. Measure and record the following values:
   V_{CEQ} = _____
   I_{CQ} = _____
   I_B = _____

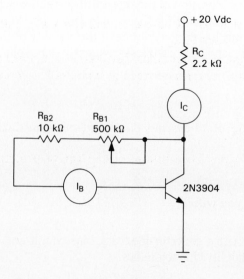

Figure 19.1

What effect, if any, has changing the transistor had on the values of V_{CEQ} and I_{CQ}?

6. If necessary, adjust the setting of R_{B1} to once again obtain midpoint bias.

Part II. Fault Symptoms

7. With power applied, remove R_C from the circuit.

> Note: Whenever you are directed to remove a component, a gap should be left where the component appeared in the circuit. Do not bridge the gap left by the component unless directed to do so.

8. Measure and record the following values:

V_{CEQ} = _____

I_{CQ} = _____

I_B = _____

9. Return R_C to its original position and remove R_B (R_{B1} and R_{B2}) from the circuit. Measure and record the following values:

V_{CEQ} = _____

I_{CQ} = _____

I_B = _____

THE BRAIN DRAIN (Optional)

10. Using circuit measurements, verify the validity of the following statement:

    A collector-feedback bias circuit will be midpoint biased when R_B is approximately equal to $h_{FE}R_C$.

11. In a separate report, include the steps you took to verify the statement, and the % of error between the values of h_{FE} found by I_C/I_B and R_B/R_C.

QUESTIONS/PROBLEMS

1. Explain your observations in step 5 of the procedure.

2. Explain the readings you obtained in step 8 of the procedure.

3. Explain the readings you obtained in step 9 of the procedure.

4. Discuss, in your own words, what you observed in this exercise.

Exercise 20

Emitter-Feedback Bias

OBJECTIVES

- To demonstrate the dc operating characteristics of the emitter-feedback bias circuit.
- To demonstrate the common fault symptoms that may occur in an emitter-feedback bias circuit.

DISCUSSION

The emitter-feedback bias circuit is a variation on the base bias circuit. It uses an added emitter resistor to reduce the shift in Q-point that can occur when h_{FE} and/or temperature changes. While this circuit is more stable against changes in h_{FE} and temperature than base bias, it still does not have the stability of voltage-divider bias.

LAB PREPARATION

Review Section 6.5 of *Introductory Electronic Devices and Circuits*.

LAB OVERVIEW

In this exercise, you will:

1. Construct an emitter feedback bias circuit and measure the midpoint bias values of V_{CEQ}, V_C, V_E, V_B, I_C, and I_B.
2. Observe the effects of a change in h_{FE} on the values listed above.
3. Observe the effects of an increase in R_E on the stability of the circuit.

4. Introduce a series of faults into the circuit and observe the effects that these faults have on the values of V_B, V_C, and V_E.

MATERIALS

1   Variable dc power supply
3   VOMs and/or DMMs
1   2N3904 npn transistor
1   2N2222 npn transistor
4   Resistors: 510 Ω, 1.5 kΩ (2), and 10 kΩ
1   500 kΩ potentiometer

> Note: Any combination of VOMs and/or DMMs is acceptable.

PROCEDURE

1. Construct the circuit shown in Figure 20.1.
2. Apply power to the circuit and adjust R_{B1} to obtain midpoint bias. Measure and record the following values:

$$V_{CEQ} = \underline{\hspace{3cm}}$$
$$V_C = \underline{\hspace{3cm}}$$
$$V_E = \underline{\hspace{3cm}}$$
$$V_B = \underline{\hspace{3cm}}$$
$$I_B = \underline{\hspace{3cm}}$$
$$I_{CQ} = \underline{\hspace{3cm}}$$

3. Using the measured values of I_{CQ} and I_B, calculate the value of h_{FE} at the Q-point.

$$h_{FE} = \underline{\hspace{3cm}}$$

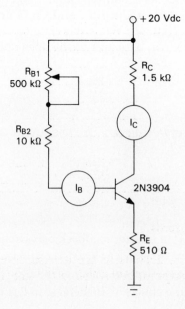

Figure 20.1

Part II. Bias Stability

4. Remove power from the circuit and replace the 2N3904 with your 2N2222 npn transistor.

> Note: Do not adjust the setting of R_{B1} until directed to do so.

5. Reapply power to the circuit and measure the following values:

V_{CEQ} = _____

V_C = _____

V_E = _____

V_B = _____

I_{CQ} = _____

I_B = _____

6. Using the measured values of I_{CQ} and I_B, calculate the value of h_{FE} at the Q-point.

h_{FE} = _____

7. Using the values of V_{CEQ} measured in steps 2 and 5, calculate the change in V_{CEQ} that occurred when you changed transistors.

$\triangle V_{CEQ}$ = _____

8. Using the values of I_{CQ} measured in steps 2 and 5, calculate the change in I_{CQ} that occured when you changed transistors.

$\triangle I_{CQ}$ = _____

Part III. Increasing The Value of R_E

9. Disconnect power from the circuit and replace R_E with your extra 1.5 kΩ resistor. Also, return the 2N3904 to the circuit.

10. Repeat steps 2 through 8 and record the measured values in the appropriate spaces.

	2N3904	2N2222
V_{CEQ}	_____	_____
V_C	_____	_____
V_E	_____	_____
V_B	_____	_____
I_B	_____	_____
I_{CQ}	_____	_____
$\triangle V_{CEQ}$ =	_____	
$\triangle I_{CQ}$ =	_____	

Part IV. Fault Symptoms

11. Return the 2N3904 to the circuit, apply power, and adjust R_{B1} to obtain midpoint bias. With power applied, remove R_C from the circuit.

> Note: Whenever you are directed to remove a component, a gap should be left where the component appeared in the circuit. Do not bridge the gap left by the component unless directed to do so.

12. Measure and record the following values:

$V_B = $ _____

$V_E = $ _____

$V_C = $ _____

13. Return R_C to its original position in the circuit and remove R_E. Measure and record the following values:

$V_B = $ _____

$V_E = $ _____

$V_C = $ _____

14. Return R_E to its original position in the circuit and remove R_B (R_{B1} and R_{B2}). Measure and record the following values:

$V_B = $ _____

$V_E = $ _____

$V_C = $ _____

THE BRAIN DRAIN (Optional)

15. Plot the dc load line for the circuit in Figure 20.1. Then, using at least 5 sets of circuit measurements, verify the validity of your load line.

16. In a separate report, show your load line, your predicted and measured values of V_{CEQ} and I_{CQ}, and an explanation of any points that fail to fall on the load line.

QUESTION/PROBLEMS

1. The percents of change in V_{CEQ} for the two circuits in this exercise are found as

$$\% \text{ change} = \frac{\Delta V_{CEQ}}{V_{CEQ1}} \times 100$$

where

$$V_{CEQ1} = \text{the original value of } V_{CEQ}$$

Using this equation, calculate the % of change that occured in V_{CEQ} when $R_E = 510 \Omega$ and when $R_E = 1.5 \text{ k}\Omega$.

% of change = _____ when $R_E = 510 \ \Omega$

% of change = _____ when $R_E = 1.5 \text{ k}\Omega$

2. Did increasing R_E reduce the % of change in V_{CEQ}? Why or why not?

3. What is your opinion of using higher values of R_E as a means of stabilizing the Q-point of an emitter-feedback bias circuit?

4. Explain the values measured in step 12 of the procedure.

5. Explain the values measured in step 13 of the procedure.

6. Explain the values measured in step 14 of the procedure.

7. Discuss, in your own words, what you observed in this exercise.

Exercise 21

BJT Testing

OBJECTIVE

- To demonstrate a practical method for testing BJTs.

DISCUSSION

For ohmmeter testing purposes, an npn transistor is similar to two diodes connected back-to-back as shown in Figure 21.1. There are two junctions: the emitter-base junction and the collector-base junction. Each of these junctions can be viewed as a diode.

When testing a diode, the junction is forward biased by the ohmmeter, and if the diode is good, a relatively low resistance is measured. Then, the ohmmeter

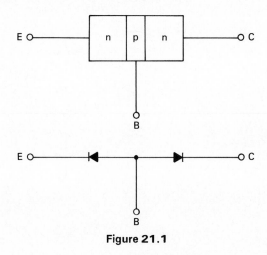

Figure 21.1

121

is connected to reverse bias the junction. If the junction is good, a relatively high resistance is measured.

When testing a transistor, forward and reverse resistance checks are performed on each of the junctions. If these readings indicate that the device is good, one more resistance reading is taken. This reading is taken between the collector and emitter terminals, and should have a relatively high value regardless of the meter polarity. If the transistor fails any of these tests, it is faulty and must be replaced.

The pnp transistor can be tested in a similar fashion. You need only remember that the ohmmeter leads must be reversed to obtain the desired readings.

LAB PREPARATION

Review section 5.7 of *Introductory Electronic Devices and Circuits*.

> Note: Be sure to review the precautions in Chapter 2 that are referenced in the above textbook reference.

LAB OVERVIEW

In this exercise, you will:

1. Perform a series of resistance tests on an npn transistor.
2. Perform a series of resistance tests on a pnp transistor.

MATERIALS

1   VOM or DMM
1   2N3904 npn transistor
1   2N3906 pnp transistor

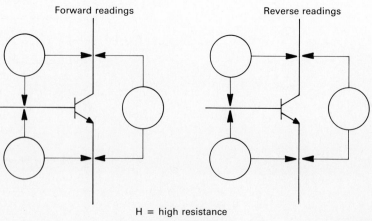

H = high resistance
L = low resistance

Figure 21.2

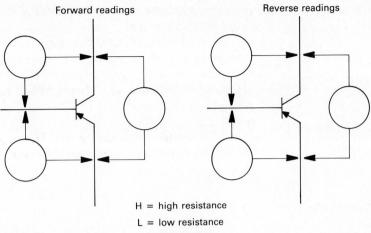

H = high resistance
L = low resistance

Figure 21.3

PROCEDURE

1. Set your VOM or DMM for resistance measurements.
2. Measure and record the junction resistance values for the npn transistor in the circles provided in Figure 21.2.
3. Repeat step 2 for the 2N3906. Record the readings in the spaces provided in Figure 21.3.

QUESTIONS/PROBLEMS

1. Discuss, in your own words, what you observed in this exercise.

Loaded Amplifiers and Swamping

OBJECTIVES

- To demonstrate the effects of *loading* on a CE amplifier.
- To demonstrate the effects of *swamping* on a CE amplifier.

DISCUSSION

The term *loading* is used to describe the addition of a load resistance to the output of an amplifier. The load resistance may be in the form of another amplifier, a speaker, or some other device that is being driven by the amplifier.

When a load is added to an amplifier, the ac collector resistance consists of the collector resistor (R_C) in parallel with the input resistance of the load. This parallel equivalent resistance *must* be less than the value of R_C. Thus, the *loaded voltage gain (A_{vL})* will always be less than the unloaded voltage gain (A_v). The actual reduction in gain depends on the value of the load resistance.

The term *swamping* is used to describe the addition of a small *unbypassed* resistance of the emitter circuit of a CE amplifier. This resistance helps to stabilize the voltage gain of the amplifier. However, as a drawback, the A_{vL} of the amplifier is significantly reduced when swamping is used. Therefore, when an amplifier is swamped, you get a more stable, but lower, value of voltage gain.

LAB PREPARATION

Review sections 7.5 and 7.6 of *Introductory Electronic Devices and Circuits*.

LAB OVERVIEW

In this exercise, you will:

Bℓ Bʀ R O y G B V Gw
0 1 2 3 4 5 6 7 8 9

132

1. Construct a CE amplifier and measure its unloaded voltage gain.
2. Add a load to the circuit and observe the effects of the load on the voltage gain of the circuit.
3. Modify the circuit so that it contains a swamping (unbypassed) emitter resistor and observe the effects that this resistance has on the voltage gain of the circuit.

MATERIALS

1 Variable dc power supply
1 Variable ac signal generator
1 Dual trace oscilloscope
1 VOM or DMM
1 2N3904 npn transistor
7 Resistors: 91 Ω, 910 Ω, 1 kΩ (2), 2.2 kΩ, 3.3 kΩ, and 3.6 kΩ
1 50 kΩ potentiometer
2 10 μF electrolytic capacitors
1 100 μF electrolytic capacitor
1 2N2222 npn transistor (optional)

PROCEDURE

3.3= Oʀ, Oʀ, Rℓ
2.2= Red, Rℓ; Bℓ
3.6k= Oʀ, Bℓ, Red
1k= Bʀ Bℓ nd

1. Construct the circuit shown in Figure 23.1.
2. Apply power to the circuit and adjust R_{1a} to provide midpoint bias.
3. Apply a 10 kHz, 20 mV$_{pp}$ ac signal at point (A) in the circuit.
4. Measure and record the peak-to-peak voltage at point (B) in the circuit.
   V_{PP} (point B) = _____
5. Using the values obtained in steps 3 and 4, calculate the voltage gain of the amplifier.
   A_v = _____

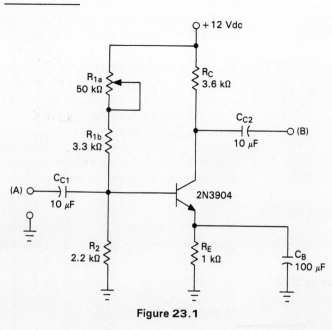

Figure 23.1

6. Disconnect the ac signal generator and add your 1 kΩ load resistor (R_L) between point (B) and ground.

7. Reconnect the signal generator to the circuit. Make sure that you still have a 10 kHz, 20 mV_{PP} input signal at point (A). Measure and record the following values:

    V_{PP} (point B) = _____

    A_{vL} = _____

Part II. Swamping

8. The circuit shown in Figure 23.2 is almost identical to the circuit you have been analyzing in this exercise. The only difference is that the 1 kΩ emitter resistor has been split into two smaller resistance values: an *unbypassed* 91 Ω resistance and a *bypassed* 910 Ω resistance. Disconnect power from the circuit and make the indicated component changes.

9. Since we have changed the total emitter resistance slightly, you will need to readjust R_{1a} to once again provide midpoint bias.

10. Using the value of A_v obtained in step 5 and the rated value of R_C, calculate the value of r_e' for the transistor.

    r_e' = _____

11. Using the value of r_e' calculated in step 10 and the rated values of r_E and R_C, predict the voltage gain of the swamped amplifier as follows:

$$A_v = \frac{R_C}{r_e' + r_E} = \text{_____}$$

12. Apply the 10 kHz, 20 mV_{PP} input signal at point (A) in the circuit. Then, measure and record the following values:

    V_{PP} (point B) = _____

    A_v = _____

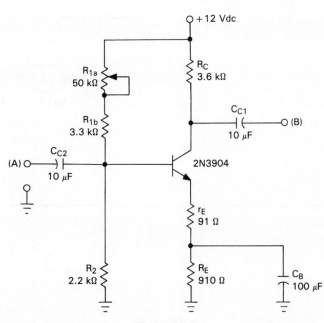

Figure 23.2

THE BRAIN DRAIN (Optional)

13. Develop a procedure to show that the circuit in Figure 23.1 will experience a larger % of change in A_v than the circuit in Figure 23.2 when the 2N3904 is replaced with a 2N2222 transistor.

14. In a separate report, include the following:
    a. your testing procedure
    b. the % of change in A_v experienced by the circuit in Figure 23.1
    c. the % of change in A_v experienced by the circuit in Figure 23.2.
    d. a brief discussion on amplifier swamping as a means of reducing variations in A_v caused by a change in transistors.

QUESTIONS/PROBLEMS

1. Step 7 of the procedure gave you a value of A_{vL} when the amplifier had a 1 kΩ load. Step 5 of the procedure gave you a value of A_v when the amplifier had an open load. Based on the measurements you made, write a simple statement about the relationship between load resistance and amplifier voltage gain.

2. What is the percent of error between the value of A_v predicted in step 11 of the procedure and the value measured in step 12?

    % of error = _____

    How would you account for this error?

3. Compare the value of voltage gain measured in step 5 of the procedure with the value measured in step 12. Based on these values of A_v, write

a simple statement about the effect of swamping on the magnitude of A_v. Use your measured values to back up your statement.

4. Discuss, in your own words, what you observed in this exercise.

Exercise 24

Cascaded CE Amplifiers

OBJECTIVE

- To demonstrate the operation of a two-stage CE amplifier.

DISCUSSION

When amplifier stages are connected in series, or *cascaded,* the overall voltage gain is much larger than either individual voltage gain. The overall voltage gain is found as the product of the individual stage voltage gain values.

Cascading is accomplished by connecting the collector of one stage to the base of the next stage. By using a coupling capacitor between the stages, the ac output of the first stage is coupled to the input of the second stage. At the same time, the coupling capacitor provides dc isolation between the stages. That is, it prevents the dc output of the first stage from affecting the dc biasing of the second.

LAB PREPARATION

Review chapter 7 of *Introductory Electronic Devices and Circuits.*

LAB OVERVIEW

In this exercise, you will:

1. Construct a two-stage CE amplifier and predict the dc terminal voltages for each stage.
2. Measure the dc terminal voltages in the amplifier.

3. Apply an ac signal to the amplifier and measure the peak-to-peak voltages at various points in the circuit.

4. Using the measured circuit voltages, determine the *total* voltage gain (A_{vT}) of the circuit.

5. Observe the effects of an added load on the value of A_{vT} for the circuit.

MATERIALS

1 Variable dc power supply
1 Variable ac signal generator
1 Dual-trace oscilloscope
1 VOM or DMM
2 2N3904 npn transistors
11 Resistors: 68 Ω (2), 1 kΩ (2), 1.5 kΩ, 2.2 kΩ(2), 3.3 kΩ (2), and 3.6 kΩ (2)
2 50 kΩ potentiometers
3 1 µF electrolytic capacitors
2 47 µF electrolytic capacitors

PROCEDURE

1. Construct the circuit shown in Figure 24.1.

2. Predict the following values for the circuit. Assume in your calculations that $R_1 = R_6 = 10\ k\Omega$.

V_{B1} = _____
V_{E1} = _____
V_{C1} = _____
V_{B2} = _____
V_{E2} = _____
V_{C2} = _____

Note: R_1 is the combination of R_{1a} and R_{1b}. R_6 is the combination of R_{6a} and R_{6b}.

3. Apply power to the circuit. Adjust R_{1a} and R_{6a} so that each of the amplifier stages is midpoint biased. Measure and record the following values:

V_{B1} = _____
V_{E1} = _____
V_{C1} = _____
V_{B2} = _____
V_{E2} = _____
V_{C2} = _____

4. Set your ac signal generator for a frequency of 10 kHz. Then, turn your amplitude down to zero and connect the generator to point (A) in the circuit.

5. Disconnect the load (R_L) from the circuit.

6. Set your oscilloscope for ac coupling and a vertical sensitivity of 2 Volts/Div. Connect the oscilloscope to point (D) in the circuit.

7. Increase the amplitude of the signal generator until you get the maxi-

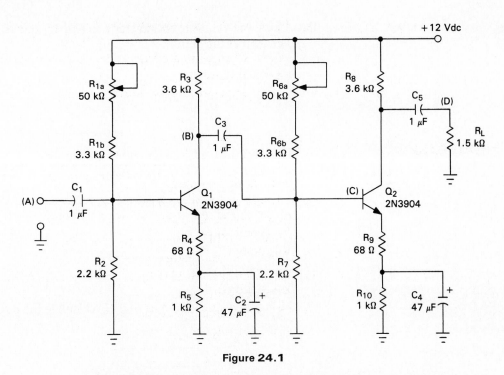

Figure 24.1

mum *undistorted* output signal at point (D) in the circuit. Record the peak-to-peak value of this output signal.

V_{PP} (point D) = _____

8. Measure and record the peak-to-peak input signal to the amplifier.

V_{PP} (point A) = _____

9. Measure and record the peak-to-peak output from the first stage.

V_{PP} (point B) = _____

10. Using the values obtained in steps 8 and 9, determine the load voltage gain of the first stage.

A_{vL1} = _____

11. Measure and record the peak-to-peak input to the second stage.

V_{PP} (point C) = _____

12. How does the value measured in step 11 compare with the one measured in step 9?

13. Using the values obtained in steps 11 and 7, determine the voltage gain of the second stage.

A_{v2} = _____

14. Using the values found in steps 10 and 13, determine the overall voltage gain of the amplifier.

$$A_{vT} = A_{vL1} A_{v2} = \underline{\hspace{3cm}}$$

15. Using the values obtained in steps 7 and 8, determine the overall voltage gain of the circuit.

$$A_{vT} = \frac{V_{out}}{V_{in}} = \underline{\hspace{3cm}}$$

16. Connect the load resistor (R_L) to the amplifier.

> Note: Do not adjust the input amplitude to the circuit.

17. Measure and record the following values:

V_{PP} (point D) = \underline{\hspace{2.5cm}}

V_{PP} (point C) = \underline{\hspace{2.5cm}}

V_{PP} (point B) = \underline{\hspace{2.5cm}}

V_{PP} (point A) = \underline{\hspace{2.5cm}}

18. Using the values listed in step 17, calculate the individual voltage gain values.

$$A_{vL1} = \underline{\hspace{3cm}}$$

$$A_{vL2} = \underline{\hspace{3cm}}$$

19. Using the data from step 18, calculate the total voltage gain of the amplifier.

$$A_{vT} = A_{vL1} A_{vL2} = \underline{\hspace{3cm}}$$

20. Using the values of V_{PP} at points (D) and (A), calculate the value of A_{vT}.

$$A_{vT} = \frac{V_{out}}{V_{in}} = \underline{\hspace{3cm}}$$

THE BRAIN DRAIN (Optional)

21. Devise a method for measuring the power gain (A_p) of each stage of the circuit in Figure 24.1, and the value of total power gain (A_{pT}) for the overall circuit.

22. In a separate report, show your procedure and results.

QUESTIONS/PROBLEMS

1. Compare the values of A_{vT} obtained in steps 14 and 15 of the procedure. What was the percent of error between these two values?

% of error = \underline{\hspace{3cm}}

How would you account for this error?

2. Compare the V_{PP} measurements that you made at point (C) in steps 11 and 17 of the procedure. Did the addition of the load resistance have an impact on the output of the first stage?

3. Based on your answer to question 2, what is your opinion of the use of a second amplifier stage as a means of isolating the first stage from the load?

4. Discuss, in your own words, what you observed in this exercise.

Exercise 26

The Common-Base Amplifier

- To demonstrate the ac operation of the *common-base (CB)* amplifier.

DISCUSSION

The CB amplifier is the least often used of the BJT amplifiers. It has low current gain, high voltage gain, low input impedance, and relatively high output impedance.

The CB amplifier is used primarily in high-frequency applications and in low-to-high impedance matching applications. In this exercise, we will take a brief look at the operation of the CB amplifier.

LAB PREPARATION

Review section 8.4 and 8.5 of *Introductory Electronic Devices and Circuits*.

LAB OVERVIEW

In this exercise, you will:

1. Construct a CB amplifier and observe its input/output phase relationship.
2. Measure the voltage gain and input impedance of the amplifier.

MATERIALS

1 Dual-polarity variable dc power supply
1 Variable ac signal generator
1 Dual-trace oscilloscope
1 VOM or DMM
1 2N3904 npn transistor
3 Resistors: 3.3 kΩ, 5.1 kΩ, and 11 kΩ
2 Potentiometers: 1 kΩ and 10 kΩ
2 4.7 µF electrolytic capacitors
1 10 kΩ potentiometer (additional, optional)

PROCEDURE

1. Construct the circuit shown in Figure 26.1
2. Apply power to the circuit and adjust R_C to obtain midpoint bias.

> Note: Don't forget . . . an emitter bias circuit is midpoint biased when
> $$V_{CEQ} = \frac{V_{CC}}{2}.$$

3. Set the amplitude of your signal generator to its lowest setting. Set the generator frequency to approximately 10 kHz.
4. Set both channels of your oscilloscope for ac coupling at a vertical sensitivity of 5 Volts/Div. Connect channel (A) to the amplifier output and channel (B) to the amplifier input.

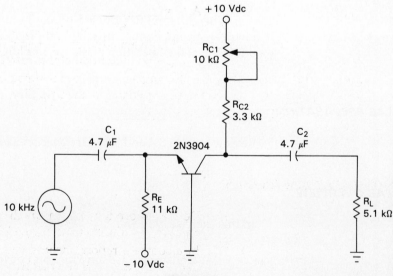

Figure 26.1

5. Increase the amplitude of the ac signal generator to obtain the maximum *undistorted* output from the amplifier.
6. Decrease the vertical sensitivity of channel (B) until you have a clear waveform displayed on the CRT.
7. Neatly draw both of the waveforms in the space provided.

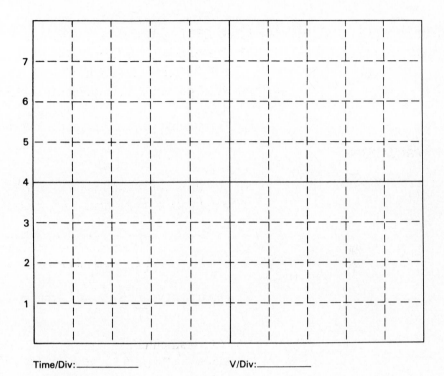

Time/Div:_____   V/Div:_____

8. From your oscilloscope display, determine the following values:

V_{out} = _____ V_{PP}

V_{in} = _____ V_{PP}

A_{vL} = _____

9. Disconnect the ac signal source from the amplifier. Insert the 1 kΩ potentiometer between the signal source and the input coupling capacitor. Set the potentiometer to its minimum setting.
10. Following the procedure outlined in step 12 of Exercise 25, measure the input impedance of the amplifier.

Z_{in} = _____

THE BRAIN DRAIN (Optional)

11. Devise a procedure to measure the current gain (A_i) and output impedance (Z_{out}) of the circuit in Figure 26.1.
12. In a separate report, include your procedure, any appropriate schematic diagrams, and the measured values of A_i and Z_{out}. Also, briefly discuss the relationship between Z_{out} and Z_{in}.

QUESTIONS/PROBLEMS

1. What phase relationship did you observe between the amplifier input and output signals?

2. How would you account for the low value of Z_{in} for the amplifier?

3. Discuss, in your own words, what you observed in this exercise.

Amplifier Compliance and Clipping

OBJECTIVES

- To demonstrate the concept of amplifier compliance (PP).
- To demonstrate the relationship between amplifier biasing and saturation and cutoff clipping.

DISCUSSION

The *compliance (PP)* of an amplifier is the maximum theoretical limit on the peak-to-peak output voltage of the circuit, found as either

$$PP = 2V_{CEQ}$$

or

$$PP = 2I_{CQ}r_C$$

When predicting the compliance of an amplifier, both of the above equations are solved for the circuit. The *lower* of the two results is the compliance of the circuit.

When an amplifier is biased below midpoint on the dc load line, the circuit may experience *cutoff clipping*. If an amplifier is biased above midpoint on the dc load line, the circuit may experience *saturation clipping*. You will see both of these types of clipping in this exercise.

LAB PREPARATION

Review section 9.2 of *Introductory Electronic Devices and Circuits*.

LAB OVERVIEW

In this exercise, you will:

1. Predict and measure the compliance of an amplifier that is midpoint biased.
2. Observe the type(s) of clipping that the midpoint biased amplifier experiences when overdriven.
3. Repeat steps 1 and 2 for an amplifier that is biased above midpoint.
4. Repeat steps 1 and 2 for an amplifier that is biased below midpoint.

MATERIALS

1 Variable dc power supply
1 Variable ac signal generator
1 Oscilloscope
1 VOM
1 DMM
1 2N3904 npn transistor

Note: You may use a second DMM in place of the VOM, if desired.

5 Resistors: 2.2 kΩ (2), 3.3 kΩ, and 10 kΩ (2)
1 50 kΩ potentiometer
2 1 μF electrolytic capacitors
1 100 μF electrolytic capacitor
3 Resistors: 4.7 kΩ, 15 kΩ, and 47 kΩ (additional-optional)

PROCEDURE

1. Construct the circuit shown in Figure 27.1.
2. Apply power to the circuit and adjust R_{1a} for midpoint bias. Measure and record the following values:

V_{CEQ} = _____

I_{CQ} = _____

3. Using the values measured in step 2 and the rated values of R_C and R_L,

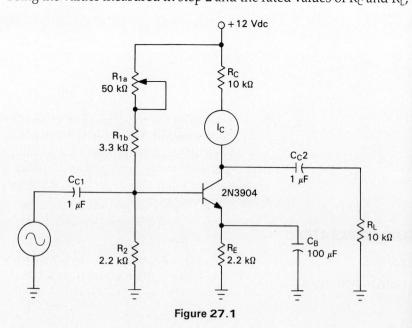

Figure 27.1

predict the compliance of the amplifier.

PP = _____

4. Apply a 10 kHz ac input to the amplifier and adjust the amplitude until you obtain the maximum *undistorted* ac load voltage. Measure and record the peak-to-peak value of the ac load signal.

V_{out} = _____ V_{PP}

5. Increase the amplitude of the ac input to the amplifier until you first observe clipping at the load. Draw the output waveform as neatly as possible in the space provided.

Time/Div:_____   V/Div:_____

6. Identify the type(s) of clipping observed in step 5.

7. Remove the ac input from the amplifier.

8. Adjust R_{1a} so that V_{CEQ} = 3 Vdc. Measure and record the following values:

V_{CEQ} = _____

I_{CQ} = _____

9. Using the values measured in step 8 and the rated values of R_C and R_L, predict the compliance of the amplifier.

PP = _____

10. Reapply the ac input to the amplifier and adjust the amplitude of the input for the maximum *undistorted* ac load voltage. Measure and record the peak-to-peak value of the ac load signal.

V_{out} = _____ V_{PP}

11. Increase the amplitude of the ac input to the amplifier until you first observe clipping at the load. Draw the output waveform as neatly as possible in the space provided.

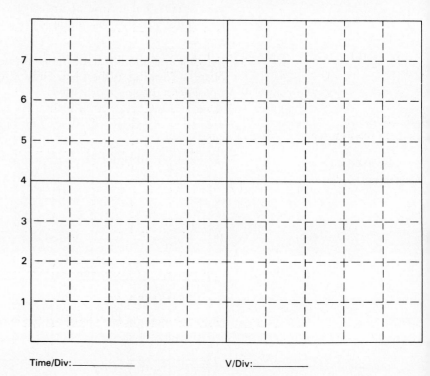

Time/Div:_____   V/Div:_____

12. Identify the type(s) of clipping observed in step 11.

13. Remove the ac input from the amplifier and adjust R_{1a} so that $V_{CEQ} = 9$ V. Measure and record the following values:

    V_{CEQ} = _____

    I_{CQ} = _____

14. Using the values measured in step 13 and the rated values of R_C and R_L, predict the compliance of the amplifier.

    PP = _____

15. Reapply the ac input to the amplifier and adjust the amplitude of the input for the maximum *undistorted* ac load voltage. Measure and record the peak-to-peak value of the ac load signal.

    V_{out} = _____V_{PP}

16. Increase the amplitude of the ac input to the amplifier until you first observe clipping at the load. Draw the output waveform as neatly as possible in the space provided on the next page.

**Step 16
Waveform**

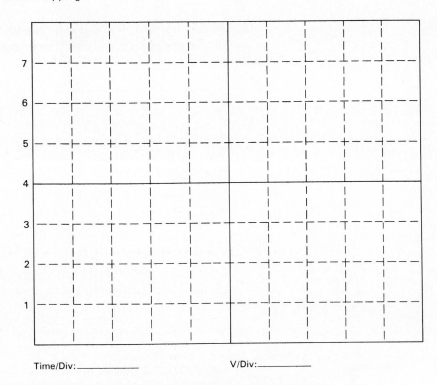

Time/Div:_____   V/Div:_____

17. Identify the type(s) of clipping observed in step 16.

THE BRAIN DRAIN (Optional)

18. Using the additional resistors listed in the *MATERIALS* section of the
     exercise, demonstrate the effects of a change in load resistance on the
     compliance and clipping of an amplifier. Maintain a constant midpoint
     bias throughout your experiment.

19. In a separate report, include your predicted and measured compliance
     values, and the types of clipping experienced by each circuit.

QUESTIONS/PROBLEMS

1. Compare the predicted and measured compliance values for the mid-
    point biased amplifier (steps 3 and 4 of the procedure). What is the
    percent of error between the two values?

    % of error = _____

    How would you account for this error?

2. In your opinion, why did you observe the type(s) of clipping identified for the midpoint based amplifier in step 6 of the procedure?

3. Compare the predicted and measured compliance values for the amplifier that is biased above midpoint (steps 9 and 10 of the procedure). What is the percent of error between the two values?

   % of error = _____

   How would you account for this error?

4. In your opinion, why did you observe the type(s) of clipping identified for the amplifier in step 12 of the procedure?

5. Compare the predicted and measured compliance values for the amplifier that is biased below midpoint (steps 14 and 15 of the procedure. What is the percent of error between the two values?

   % of error = _____

   How would you account for this error?

6. In your opinion, why did you observe the type(s) of clipping identified for the amplifier in step 17 of the procedure?

7. Discuss, in your own words, what you observed in this exercise.

Class AB Amplifiers

OBJECTIVES

- To demonstrate the dc and ac operating characteristics of the class AB complementary-symmetry amplifier.
- To demonstrate some of the common fault symptoms that may occur in a class AB amplifier.

DISCUSSION

The class AB amplifier uses two transistors that each conduct for slightly more than 180° of the ac input cycle to produce a linear output without crossover distortion.

Since the two transistors are connected in series in a class AB amplifier, the values of I_{CQ} for the two devices are nearly equal. Since both transistors are biased just above cutoff, the overall value of I_{CQ} for the amplifier is relatively low. This results in a relatively high amplifier efficiency; theoretically up to about 78.5%.

LAB PREPARATION

Review sections 9.5 through 9.7 of *Introductory Electronic Devices and Circuits*.

LAB OVERVIEW

In this exercise, you will:

1. Construct a class AB amplifier and measure the dc voltages in the circuit.
2. Measure the voltage gain, compliance, and efficiency of the circuit.
3. Introduce a series of faults into the amplifier and observe their symptoms.

MATERIALS

1 Variable dc power supply
1 Variable ac signal generator
1 Oscilloscope
1 VOM or DMM
1 2N3904 npn transistor
1 2N3906 pnp transistor
2 1N4148 small-signal diodes
2 Resistors: 27 Ω and 2 kΩ
1 5 kΩ potentiometer
2 10 μF electrolytic capacitors
1 470 μF electrolytic capacitor
2 resistors: 120Ω and 1kΩ (additional, optional)

PROCEDURE

1. Construct the circuit shown in Figure 28.1. The value of R_1 should initially be set to approximately 2 kΩ.
2. Apply power to the circuit and adjust R_1 to provide midpoint bias. Measure and record the following:

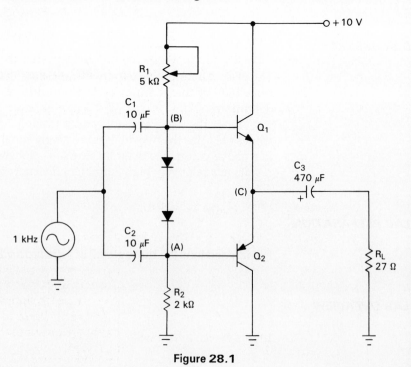

Figure 28.1

V_{CEQ1} = _____ $V_{B(Q1)}$ = _____
V_{CEQ2} = _____ $V_{B(Q2)}$ = _____

3. Set the frequency of your ac signal generator to approximately 1 kHz. Reduce its output amplitude to its minimum setting.

4. Apply the input signal to the amplifier. Then, increase the amplitude of the input signal until you obtain the maximum *undistorted* output from the amplifier. Measure and record the peak-to-peak output from the circuit.

    V_{out} = PP = _____V_{PP}

5. Measure and record the peak-to-peak input to the amplifier.

    V_{in} = _____V_{PP}

6. Using the values obtained in steps 4 and 5, calculate the voltage gain of the amplifier.

    A_{vL} = _____

7. Using the value obtained in step 4, calculate the value of $I_{C(sat)}$ for the amplifier, as follows:

$$I_{C(sat)} = \frac{V_{PP}}{2R_L}$$

    $I_{C(sat)}$ = _____

8. Using the value of $I_{C(sat)}$ obtained in step 7, calculate the value of $I_{C(ave)}$ as follows:

$$I_{C(ave)} = \frac{I_{C(sat)}}{\Pi}$$

    $I_{C(ave)}$ = _____

9. Using the value of $I_{C(ave)}$ obtained in step 8, calculate the value of source power (P_S) as follows:

$$P_S = I_{C(ave)} V_{CC}$$

    P_S = _____

10. Using the value of V_{out} measured in step 4 and the rated value of R_L, calculate the load power of the amplifier.

    P_L = _____

11. Using the values calculated in steps 9 and 10, calculate the efficiency of your amplifier.

    η = _____

Part II. Fault Symptoms

12. Remove R_1 from the circuit.

> Note: Whenever you are directed to remove a component, a gap should be left where the component appeared in the circuit. Do not bridge the gap left by the component unless directed to do so.

Measure and record the voltages from the following points to ground.

$V_A =$ _____ $V_B =$ _____ $V_C =$ _____

13. Return R_1 to its original position and remove D_1. Reapply power to the circuit. Measure and record the voltages from the following points to ground.

$V_A =$ _____ $V_B =$ _____ $V_C =$ _____

14. Return D_1 to the circuit. Remove R_2 and reapply power to the circuit. Measure and record the voltages from the following points to ground.

$V_A =$ _____ $V_B =$ _____ $V_C =$ _____

THE BRAIN DRAIN (Optional)

15. Using the additional resistors listed in the MATERIALS section of the exercise, demonstrate the effects of a change in load resistance on the voltage gain and efficiency of a class AB amplifier.

16. In a separate report, include predicted and measured values of A_{vL} and efficiency.

QUESTIONS/PROBLEMS

1. The different amplifier configurations have different voltage gain characteristics. Based on your measured value of A_{vL}, what is the circuit configuration of the amplifier? Explain your answer.

2. Explain the readings you obtained in step 12.

3. Explain the readings you obtained in step 13.

4. Explain the readings you obtained in step 14.

5. Discuss, in your own words, what you observed in this exercise.

An Audio Amplifier

OBJECTIVES

- To demonstrate the operation of a two stage audio amplifier that utilizes a class AB amplifier.
- To demonstrate another method of coupling a signal to a class AB amplifier.

DISCUSSION

The basic audio amplifier (Figure 29.1) consists of a voltage amplifier (Q_1), a driver amplifier (Q_2), and a class AB complementary-symmetry amplifier. In the circuit shown, R_7 is used to set the dc output voltage to midpoint, allowing the maximum unclipped output ac signal. A speaker is placed at the output (in place of R_L) responds to any changes in input frequency.

LAB PREPARATION

Review Chapter 9 of *Introductory Electronic Devices and Circuits,* along with your results from exercise 28 in this manual.

LAB OVERVIEW

In this exercise you will:

1. Construct the circuit in Figure 29.1 and adjust it for midpoint bias.

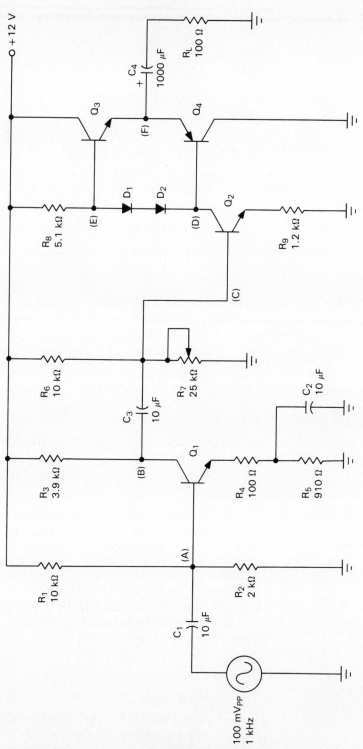

Figure 29.1

2. Measure and record the dc voltages throughout the circuit.
3. Measure the voltage gain of each amplifier stage.
4. Replace the load resistor with a speaker and observe the effects of a changing input signal on the output from the speaker.

MATERIALS

1   Variable dc power supply
1   Variable ac signal generator
1   Oscilloscope
1   VOM or DMM
3   2N3904 npn transistors
1   2N3906 pnp transistor
2   1N4148 small-signal diodes
9   Resistors: 100 Ω (2), 910 Ω, 1.2 kΩ, 2 kΩ, 3.9 kΩ, 5.1 kΩ, 10 kΩ (2)
1   25 kΩ potentiometer
3   10 μF electrolytic capacitor
1   1000 μF electrolytic capacitor
1   8 Ω speaker

PROCEDURE

1. Construct the circuit shown in Figure 29.1.
2. Before applying the ac input signal, measure the dc voltage from point (F) to ground. Adjust R_7 until this voltage is 6 Vdc.
3. Measure and record the following dc voltages:

    $V_{B(Q1)}$ = _____ $V_{B(Q3)}$ = _____

    $V_{C(Q1)}$ = _____ $V_{C(Q3)}$ = _____

    $V_{E(Q1)}$ = _____ $V_{E(Q1)}$ = _____

    $V_{B(Q2)}$ = _____ $V_{B(Q4)}$ = _____

    $V_{C(Q2)}$ = _____ $V_{C(Q4)}$ = _____

    $V_{E(Q2)}$ = _____ $V_{E(Q4)}$ = _____

4. Set the ac signal generator for a 100 mV$_{PP}$ output at a frequency of 1 kHz. Apply the input signal to the amplifier. Measure and record the peak-to-peak voltages at the following points:

    V_A = _____ V$_{PP}$ V_D = _____ V$_{PP}$

    V_B = _____ V$_{PP}$ V_E = _____ V$_{PP}$

    V_C = _____ V$_{PP}$ V_F = _____ V$_{PP}$

5. Using the values obtained in step 4, calculate the following values:

    A_v (A to B)   = _____

    A_v (C to D)   = _____

    A_v (E to F)   = _____

6. Remove power from the circuit and replace R_L with your speaker.
7. Reapply power to the circuit. Vary the amplitude setting on your ac signal generator. What is the effect on the tone being generated?

8. Set the amplitude of the input signal back to approximately 100 mV$_{PP}$. Now, vary the frequency setting of the signal generator. What is the effect on the tone being generated?

THE BRAIN DRAIN (Optional)

9. Without disturbing its setting, remove R_7 from the circuit and measure its value.

10. Perform a complete mathematical dc analysis of the amplifier and compare your calculated values to the values measured in step 3 of the procedure.

11. Using the spec sheets of the 2N3904 and 2N3906, calculate the individual stage voltage gain values and compare your results to those obtained in step 5 of the procedure.

12. In a separate report, show your calculated and measured values, the percents of error between each calculated and measured value, and an explanation of those percents of error.

QUESTIONS/PROBLEMS

1. Using the values obtained in step 5 of the procedure, calculate the overall voltage gain of the audio amplifier.

   $A_{vT} = $ _____

2. Using the values of V_{PP} at points (A) and (F) in the circuit (from step 4 of the procedure), calculate the overall voltage gain of the amplifier.

   $A_{vT} = $ _____

3. What is the percent of error between the values of A_{vT} found in questions 1 and 2?

   % of error = _____

   How would you account for this percent of error?

4. Discuss, in your own words, what you observed in this exercise.

PART IV

FETs and FET Circuits

Exercise 30

JFET Operation

OBJECTIVES

- To demonstrate the effect of drain-source voltage (V_{DS}) on drain current (I_D) when the JFET is operated with zero gate-source bias (V_{GS}).
- To demonstrate what happens when V_{DS} exceeds the pinchoff voltage (V_P) rating of a given JFET.

DISCUSSION

In terms of V_{DS}, a JFET has two distinct regions of operation. The value of V_{DS} that separates these two regions is called the pinchoff voltage (V_P). As long as V_{DS} is less than V_P, changes in V_{DS} will cause proportional changes in drain current (I_D). If V_{DS} increases, so will I_D. If V_{DS} decreases, so will I_D. Both of these statements assume, of course, that the JFET input voltage (V_{GS}) is held constant.

When V_{DS} increases above the value of V_P, further increases in V_{DS} do not affect I_D. In other words, the JFET drain becomes a constant-current circuit. Again, this assumes that there is no change in V_{GS}.

In this exercise, we will take a look at the effect of V_{DS} on I_D both above and below the value of V_P.

LAB PREPARATION

Review section 10.1 of *Introductory Electronic Devices and Circuits*.

LAB OVERVIEW

In this exercise, you will:

1. Obtain the rated values of V_P, I_{DSS}, and BV_{DSO} from the spec sheet for your JFET.
2. Take a series of V_{DS} versus I_D measurements for your JFET.
3. Using the measured V_{DS}/I_D combinations, plot the drain curve for your JFET with $V_{GS} = 0$ V.
4. From your curve, determine the values of V_P and I_{DSS} and compare the actual values to the rated values obtained from the spec sheet.

MATERIALS

1   Dual-polarity variable dc power supply
2   VOMS and/or DMMs
1   2N5485, along with its specification sheet.
1   100 Ω resistor

> Note: At least one of the meters should be a DMM. If one of your meters is a VOM, use it to measure I_D.

PROCEDURE

1. Construct the circuit shown in Figure 30.1.
2. When $V_{GS} = 0$ V (as is the case in Figure 30.1), V_P is the positive equivalent of $V_{GS(off)}$. Using your spec sheet, determine the rated range of V_P for your JFET.

   $V_P = $ _____ to _____
3. Using your spec sheet, determine the range of I_{DSS} for your JFET.

   $I_{DSS} = $ _____ to _____
4. As with any component, the JFET will break down if the voltage across its output terminals exceeds the breakdown voltage rating for those

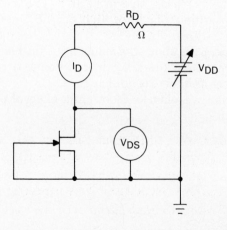

Figure 30.1

terminals. From the spec sheet, determine the drain-source breakdown voltage rating for your component. Record this rating below.

BV_{DSO} = _____

> Note: Make sure that you do not allow V_{DS} to approach the value of BV_{DSO}.

5. Apply power to the circuit and adjust V_{DD} so that V_{DS} is as close as possible to +0.5 Vdc. Record the value of I_D in the appropriate space in Table 1.

TABLE 1

V_{DS} (volts)	I_D (mA)	V_{DS} (volts)	I_D (mA)
0.0	_____	4.0	_____
0.5	_____	4.5	_____
1.0	_____	5.0	_____
1.5	_____	6.0	_____
2.0	_____	7.0	_____
2.5	_____	8.0	_____
3.0	_____	10.0	_____
3.5	_____	15.0	_____

6. For each value of V_{DS} in Table 1, measure and record the value of I_D.
7. Plots the points from Table 1 on the graph in Figure 30.2. Connect the points with a smooth curve. Your curve should be similar to the one shown in Figure 10.7a on page 423 of the text.
8. On your curve, mark the point where the sharp increase in current stops and the value of I_D levels off. Draw a vertical line down from this point to the x-axis of the graph. Read the pinchoff voltage at this point.

V_P = _____

9. The gate-source junction was shorted throughout the procedure. Therefore, by definition, the value of I_D is I_{DSS} when V_{DS} reaches V_P. Record the value of I_{DSS} (obtained from your graph).

I_{DSS} = _____

THE BRAIN DRAIN (Optional)

10. Using the negative dc output from your power supply, determine the value of V_P when V_{GS} = −0.5 V and −1 V.
11. In a separate report, draw the circuit you used to take your measurements, your results, and a brief discussion on the relationship between V_P and V_{GS}.

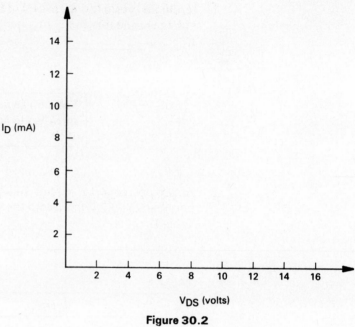

Figure 30.2

QUESTIONS/PROBLEMS

1. What is the relationship between V_{DS} and I_D when V_{DS} is less than V_P?

2. Compare your measured value of V_P with the rated range. Did it fall within the specified range?

3. Compare your measured value of I_{DSS} with the rated range. Did it fall within the specified range?

4. Discuss, in your own words, what you observed in this exercise.

Exercise 31

JFET Transconductance Curves

OBJECTIVES

- To demonstrate the effects of V_{GS} on JFET drain current.
- To provide the opportunity to plot the transconductance curve of a JFET.

DISCUSSION

The JFET is a voltage controlled device. That is, the drain current (I_D) of a given JFET is controlled by the amount of reverse bias that is applied to the gate-source junction. When V_{GS} is at its minimum value (zero), drain current is at its maximum value (I_{DSS}). Then, as V_{GS} is made to be more and more negative, the value of I_D decreases. Eventually, when V_{GS} reaches $V_{GS(off)}$, drain current drops to approximately zero.

LAB PREPARATION

Review section 10.1 of *Introductory Electronic Devices and Circuits*.

LAB OVERVIEW

In this exercise, you will:

1. Measure the values of $V_{GS(off)}$ and I_{DSS} for your JFET.
2. Take a series of I_D versus V_{GS} measurements.

3. Use the I_D/V_{GS} combinations to plot the transconductance curve for your JFET.

MATERIALS

1   Dual-polarity variable dc power supply
1   VOM
1   DMM
1   2N5485 n-channel JFET
2   Resistors: 100 Ω and 1 kΩ
1   10 kΩ potentiometer

> Note: You may use a second VOM in place of the DMM if desired.

PROCEDURE

1. Construct the circuit shown in Figure 31.1. Use a VOM for the current meter.
2. Using the following procedure, measure the value of I_{DSS} for your JFET:
    a. Set R_{G2} so that V_{GS} = 0 Vdc. Set V_{DD} to 0 Vdc.
    b. Increase V_{DD} until I_D levels off.
    c. *Slowly* decrease V_{DD} while keeping an eye on the VOM. Record the value of I_D that occurs *just before* the current drops off rapidly.
    d. Repeat the procedure two more times. Then, record the average of the three readings as the value of I_{DSS} for your JFET.

    Trial 1: _____   Trial 2: _____

    Trial 3: _____   Average: _____
3. Increase V_{DD} until I_D levels off again.
4. Increase V_{GS} until I_D reaches its lowest value. Measure and record the value of V_{GS} at which this occurs.

    $V_{GS(off)}$ = _____
5. Using your measured values of I_{DSS} and $V_{GS(off)}$ as guides, determine the appropriate I_D and V_{GS} scales to use in Figure 31.2. Be sure to provide at least four equal increments on the V_{GS} scale between 0 V and your measured value of $V_{GS(off)}$.

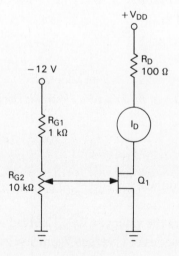

Figure 31.1

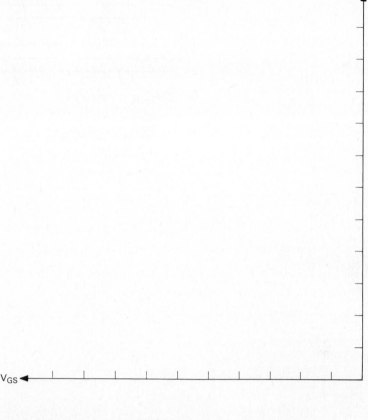

Figure 31.2

6. Plot the points that correspond to $V_{GS(off)}$ and I_{DSS} on the graph.

7. For each increment value of V_{GS} on your graph, measure and record the value of I_D. Adjust R_G in each case to obtain the desired value of V_{GS}.

$I_D = $ _____ at $V_{GS} = $ _____

$I_D = $ _____ at $V_{GS} = $ _____

$I_D = $ _____ at $V_{GS} = $ _____

$I_D = $ _____ at $V_{GS} = $ _____

(Continue on a separate sheet if necessary.)

8. Plot the points that represent the values in step 7 on your graph. Connect the points with a smooth curve. This curve is the transconductance curve for your JFET, and should resemble the one shown in Figure 10.11 on page 427 of the text.

QUESTIONS/PROBLEMS

1. Using equation (10.1) and your measured values of I_{DSS} and $V_{GS(off)}$, calculate the value of I_D for each of the V_{GS} values listed in step 7 of the procedure. Then, calculate the percent of error between your calculated and measured values. Use the table below to record your calculated and measured values.

2. How would you explain the wide variation in your percents of error in question 1?

V_{GS}	I_D (measured)	I_D (calculated)	% of error

3. Discuss, in your own words, what you observed in this exercise.

Gate Bias and Self-Bias Circuits

OBJECTIVES

- To demonstrate the dc operating characteristics of the gate bias circuit.
- To demonstrate the dc operating characteristics of the self-bias circuit.
- To demonstrate the bias instability of the gate bias and self-bias circuits.

DISCUSSION

You have seen that the drain current of a given JFET is varied by varying the amount of reverse bias applied to the gate-source junction of the device. In this exercise, you will see how two JFET biasing circuits develop the value of V_{GS} needed to establish the JFET operating point.

The gate bias circuit simply uses a negative voltage source, connected to the gate, to reverse bias the gate-source junction. The self-bias circuit is designed so that the gate is held at ground potential, while a positive voltage is developed across the source resistor. Since the gate is grounded and the source is at a positive potential, a *negative* V_{GS} exists, and the Q-point is established.

The primary drawback to using either of these biasing circuits is that they both provide extremely unstable Q-points. We will investigate this instability along with the basic dc operating principles of gate bias and self-bias circuits in this exercise.

LAB PREPARATION

Review section 10.2 of *Introductory Electronic Devices and Circuits*.

LAB OVERVIEW

In this exercise, you will:

1. Predict the Q-point limits of I_D for a gate bias circuit and then measure its actual drain current.
2. Observe the effects of a change in JFETs on the gate bias circuit.
3. Plot the minimum and maximum transconductance curves for your JFET and then plot the bias line of a self-bias circuit in order to determine the range of Q-point values.
4. Measure and record the actual values of V_{GS}, I_D, and V_{DS} for your self-bias circuit.
5. Observe the effects of a change in JFETs on the self-bias circuit.

MATERIALS

1 Dual polarity variable dc power supply
2 VOMs and/or DMMs
2 2N5485 n-channel JFETs (along with the device specification sheet)
3 Resistors: 3.3 kΩ, 4.7 kΩ, and 1 MΩ
2 Potentiometers: 25 kΩ and 10 kΩ

PROCEDURE

1. Using your spec sheet, determine the ranges of I_{DSS} and $V_{GS(off)}$ for your JFET.

   $I_{DSS} = $ _____ to _____

   $V_{GS(off)} = $ _____ to _____

2. Using the values above and equation (10.1), predict the range of I_D when $V_{GS} = -1$ Vdc.

   $I_D = $ _____ to _____ when $V_{GS} = -1$ Vdc

3. Construct the circuit shown in Figure 32.1a. R_D should initially be set to is maximum value.

4. Adjust the value of R_D until the circuit is midpoint biased. Record your measured value of V_{DS}.

   $V_{DS} = $ _____

5. Measure and record the value of I_D.

   $I_D = $ _____

6. Let your circuit sit for a moment or two. Then, check the values of I_D and V_{DS}. Have they changed? If so, describe the change.

7. If needed, adjust R_{D2} until the circuit is midpoint biased once again.

8. Remove power from the circuit. Without disturbing the setting of your

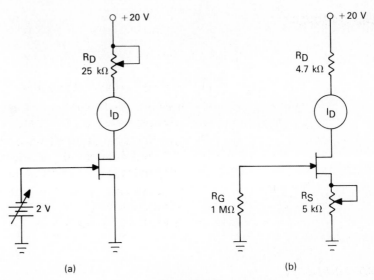

(a)   (b)

Figure 32.1

drain resistor, remove your JFET from the amplifier and replace it with the other one. Reapply power to the circuit. Measure and record the following.

V_{GS} = _____

V_{DS} = _____

I_D = _____

9. Using the values recorded in step 1, plot the maximum and minimum transconductance curves for your JFET in figure 32.2.

10. Construct the circuit in Figure 32.1b. R_S should initially be set to a value of 1kΩ

11. Plot the dc bias line for the circuit on your transconductance curves. Then, determine the ranges of V_{GS} and I_D for the circuit.

V_{GS} = _____ to _____

I_D = _____ to _____

12. Apply power to the circuit and adjust R_S to provide midpoint bias. Measure and record the following values:

V_G = _____ V_S = _____

V_{GS} = _____ I_D = _____

V_{DS} = _____

13. Remove power from the circuit. Without disturbing the setting of your source resistor, remove your JFET from the circuit and replace it with the other one. Reapply power to the circuit.

Measure and record the following:

V_{GS} = _____

I_D = _____

V_{DS} = _____

THE BRAIN DRAIN (Optional)

14. Disconnect power from your circuit. Do not disturb the setting of R_S.

15. Using the values recorded in step 13, calculate the value of R_S.

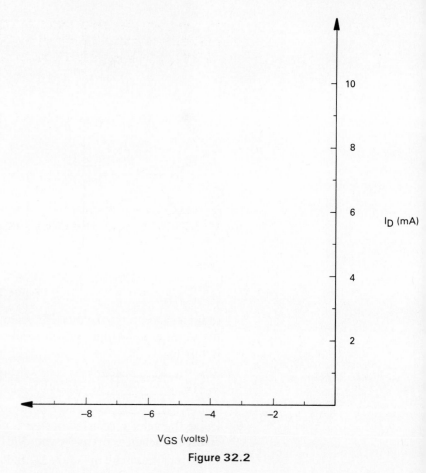

Figure 32.2

16. Remove R_S from the circuit and measure its value.
17. In a separate report, show your calculations, the predicted and measured values of R_S, the percent of error between the two values, and an explanation for the error.

QUESTIONS/PROBLEMS

1. Plot the point that represents the I_D/V_{GS} combination in step 12 of the procedure on your graph from step 9. Does the point fall on the dc bias line? If not, how would you explain the fact that it doesn't?

2. Plot the point that represents the I_D/V_{GS} combination in step 13 of the procedure on your graph from step 9. Does the point fall on the dc bias line? If not, how would you explain the fact that it doesn't?

3. Discuss, in your own words, what you observed in this exercise.

Exercise 33

Small-Signal CS Amplifiers

- To demonstrate the ac operation of a typical common-source amplifier.
- To demonstrate the differences between the ac operating characteristics of a typical JFET amplifier and those of a typical BJT amplifier.

DISCUSSION

There are several similarities between the common-source (CS) amplifier and the common-emitter (CE) amplifier. Both amplifiers provide a measurable amount of voltage gain. Both amplifiers have a 180° voltage phase shift between the input and output terminals. At the same time, there are several differences between the two amplifier types.

The CS amplifier typically has much higher input impedance than the CE amplifier. Also, the means by which the voltage gain of a CS amplifier is calculated is different that the CE voltage gain calculation.

In this exercise, we will observe the operation of a self-biased CS amplifier. While analyzing this operation, we will pay close attention to those points of operation that separate the CS amplifier from the CE amplifier.

LAB PREPARATION

Review section 10.3 of *Introductory Electronic Devices and Circuits*.

LAB OVERVIEW

In this exercise, you will:

1. Measure the value of transconductance (g_m) for your JFET.
2. Predict the value of A_v for a self-biased CS amplifier, using the measured value of g_m and the rated value of R_D.
3. Measure the A_v of the amplifier.
4. Observe the input/output phase relationship of the circuit.
5. Measure the input impedance of the circuit.

MATERIALS

1 Variable dc power supply
1 Variable ac signal generator
1 Dual trace oscilloscope
1 VOM
1 DMM
1 2N5485 n-channel JFET
2 Resistors: 4.7 kΩ and 1 MΩ
2 Potentiometers: 5 kΩ and 2.5 MΩ
2 Capacitors: 0.022 µF and 22 µF

Note: You may use a second DMM in place of the VOM if desired.

PROCEDURE

1. Construct the circuit in Figure 33.1. R_S should initially be set to 1 kΩ.
2. The value of g_m for a JFET is found as

$$g_m = \frac{\Delta I_D}{\Delta V_{GS}}$$

In this step, you will determine the value of g_m for your JFET as follows:
a. With R_S set to 1 kΩ, measure and record the following:
V_{GS} = _____
I_D = _____
b. Adjust R_S to a value of 1.5 kΩ. Then, return it to the circuit and measure and record the following:
V_{GS} = _____
I_D = _____
c. Calculate the following:
ΔV_{GS} = $V_{GS(max)} - V_{GS(min)}$ = _____
ΔI_D = $I_{D(max)} - I_{D(min)}$ = _____
g_m = $\dfrac{\Delta I_D}{\Delta V_{GS}}$ = _____

3. Using the rated value of R_D, calculate the voltage gain of your amplifier.
A_v = _____

4. Set R_S to approximately 1.3 kΩ.

5. Set the amplitude of your signal generator to minimum. Set the output frequency of your signal generator to approximately 1 kHz.

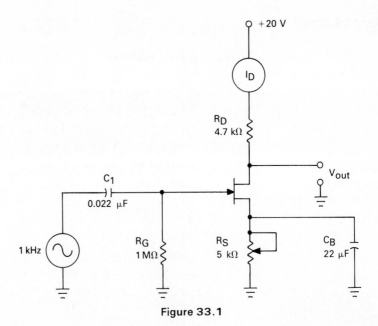

Figure 33.1

6. Connect your oscilloscope to the output of the amplifier. Then, increase the amplitude of the input signal until you get the maximum undistorted output from the amplifier. Measure and record the following:

$$V_{out} = \underline{\hspace{2cm}} V_{PP}$$
$$V_{in} = \underline{\hspace{2cm}} V_{PP}$$

7. Using the values obtained in step 6, calculate the value of A_v for your amplifier.

$$A_v = \underline{\hspace{2cm}}$$

8. Adjust the vertical sensitivity of your oscilloscope channels so that you can observe both the input and output waveforms on the CRT at the same time. Then, neatly draw the two waveforms in their proper phase relationship in the space provided on the next page.

9. Without disturbing the amplitude setting of your signal generator, insert the 2.5 MΩ potentiometer between the generator output and the input coupling capacitor.

10. Adjust the 2.5 MΩ potentiometer until the output from the amplifier has an amplitude that is one-half of its original value.

11. Without disturbing the potentiometer setting, remove it from the circuit and measure its resistance. This resistance is approximately equal to the input impedance of the amplifier.

$$R \cong Z_{in} = \underline{\hspace{2cm}}$$

THE BRAIN DRAIN (Optional)

12. Using the spec sheet for your JFET, calculate the worst-case values of A_v for the amplifier in Figure 33.1. Then verify your results by measuring A_v for the circuit using at least three different JFETs.

13. In a separate report, include your calculations and your measured

Step 8
Waveform

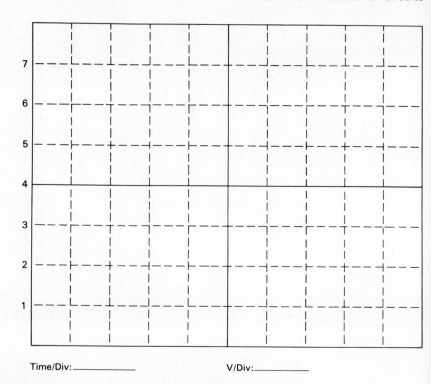

Time/Div:_____   V/Div:_____

values of A_V. Also, discuss briefly your opinion on the validity of using a worst-case analysis to predict circuit behavior.

QUESTIONS/PROBLEMS

1. Calculate the percent of error between the values of A_v obtained in steps 3 and 7 of the procedure.
   % of error = _____
   How would you account for this error?

2. Calculate the percent of error between the measured input impedance of the amplifier and the rated value of R_G.
   % of error = _____

Why wasn't the input impedance of the JFET considered in the percent of error calculation?

3. Discuss, in your own words, what you observed in this exercise.

Testing JFETs

OBJECTIVE

- To demonstrate a practical method of testing JFETs with an ohmmeter.

DISCUSSION

For ohmmeter testing purposes, the n-channel JFET can be viewed as the equivalent circuit shown in Figure 34.1b. The diode represents the gate-source junction of the JFET, while the resistance represents the ohmic resistance of the JFET channel.

When reverse biased, the gate-source junction of the JFET should have extremely high resistance. This high resistance can be measured between the gate

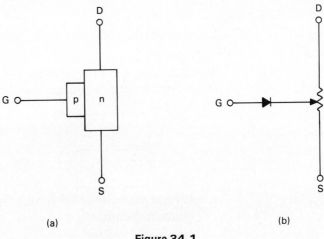

(a)   (b)

Figure 34.1

and either the source or the drain. The resistance between the source and drain should be relatively low regardless of the meter polarity.

LAB PREPARATION

Review the material on ohmmeter diode testing in section 2.10 of *Introductory Electronic Devices and Circuits*.

LAB OVERVIEW

In this section, you will:

1. Measure the reverse resistance of the gate-source and gate-drain junctions of your JFET.
2. Measure the forward resistance of the gate-source and gate-drain junctions of your JFET.
3. Measure the resistance of the channel of your JFET.

MATERIALS

1   VOM or DMM
1   2N5485 n-channel JFET
1   Variable dc power supply (optional)
1   100 kΩ resistor (optional)

PROCEDURE

1. Observing the proper meter polarity, measure the resistance of the *reverse biased* gate-source junction.

   R_{G-S} = _____

   > Note: If the resistance of the junction is too high to obtain a reading on your meter, simply indicate that its resistance is infinite.

2. Measure the resistance of the *reverse biased* gate-drain junction.

   R_{G-D} = _____

3. Observing the proper meter polarity, measure the resistance of the *forward biased* gate-source junction.

   R_{G-S} = _____

4. Measure the resistance of the *forward biased* gate-drain junction.

   R_{G-D} = _____

5. Measure the resistance from the source to the drain.

   R_{S-D} = _____

6. Reverse your meter leads and remeasure the resistance between the source and drain.

   R_{S-D} = _____

THE BRAIN DRAIN (Optional)

7. Devise a method for measuring the reverse resistance of the gate-source and gate-drain junctions using a dc power supply and a 100 kΩ resistor.

> Hint: You will need to use a
> DMM to take your voltage
> measurements. Also, be
> aware of the reverse break-
> down ratings for your JFET.

8. In a separate report, include your procedure, a schematic diagram of you circuit, your readings, and calculated junction resistances. Also, if you were able to obtain readings in steps 1 and 2 of the procedure, include a comparison of the resistance values obtained with your procedure and the previous readings.

QUESTIONS/PROBLEMS

1. How do the results in steps 1 and 2 of the procedure compare with each other?

2. How do the readings in steps 3 and 4 of the procedure compare with each other?

3. How do the readings in steps 5 and 6 of the procedure compare with each other?

4. In step 3 of the procedure, you measured the forward resistance of the gate-source junction. Given the fact that the JFET is always operated with this junction reverse biased, was this measurement really necessary? Explain your answer.

5. While answering questions 1, 2, and 3, you should have observed that the readings listed were usually very close to each other. With this in mind, what is the minimum number of measurements needed to completely test a JFET? Explain your answer.

6. Discuss, in your own words, what you observed in this exercise.

PART V
Differential Amplifiers
and Basic Op-Amp Circuits

The Differential Amplifier

OBJECTIVES

- To demonstrate the operation of the discrete component differential amplifier.
- To demonstrate the origins of some of the op-amp parameters.

DISCUSSION

The input circuit of an op-amp is a differential amplifier. A differential amplifier is a circuit that is designed to amplify the difference between two input signals. When the two input signals to a differential amplifier are equal, the amplifier actually attentuates the signals. That is, the output from the circuit actually has a much lower amplitude than the input signals.

LAB PREPARATION

Review section 12.3 of *Introductory Electronic Devices and Circuits*.

LAB OVERVIEW

In this exercise, you will:

1. Construct a discrete differential amplifier and adjust the circuit to eliminate its output offset voltage.

2. Measure its input offset voltage, input offset current, and input bias current.

3. Measure its differential gain and common-mode gain.

4. Calculate the CMRR of the circuit using the measured gain values.

MATERIALS

1   Dual-polarity variable dc power supply
1   Variable ac signal generator
1   VOM or DMM
1   Dual trace oscilloscope
2   3N3904 npn transistors
5   Resistors: 2.7 kΩ (2), 4.7 kΩ (2), 10 kΩ
2   Potentiometers: 1 kΩ and 10 kΩ

PROCEDURE

1. Construct the circuit shown in Figure 35.1.

2. Use jumper wires to connect the two transistor base terminals to ground.

3. Connect a voltmeter across the two collector terminals and adjust R_6 until the voltmeter reads 0 Vdc. This eliminates any imbalance in the amplifier.

4. Remove the jumper wires from the circuit. Measure and record the following voltages.

   $V_{B(Q1)}$ = _____ $V_{B(Q2)}$ = _____

5. Calculate the difference between the two V_B readings. This is the circuit's input offset voltage.

   Input offset voltage = _____

6. Measure and record the following currents.

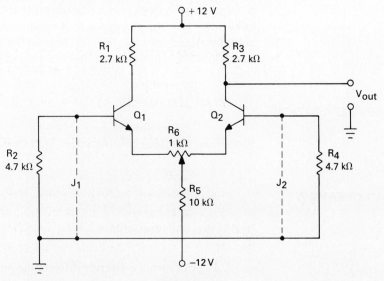

Figure 35.1

$I_{B(Q1)}$ = _____ $I_{B(Q2)}$ = _____

7. Calculate the difference between the two I_B readings. This is the circuit's input offset current.

    Input offset current = _____

8. Calculate the average of the two values of I_B. This is the circuit's input bias current rating.

    Input bias current = _____

9. Adjust your signal generator for a 400 mV$_{PP}$ output at a frequency of approximately 1 kHz.

10. Apply the input signal to the base of Q_1.

11. Using your oscilloscope, observe the input and output signals simultaneously. Draw the waveforms in the space provided.

Time/Div:_____   V/Div:_____

12. Disconnect the signal generator and apply the input signal to the base of Q_2. Draw the input and output waveforms in the space provided on the next page.

13. Measure and record the input and output peak-to-peak voltages.

    V_{in} = _____ V_{PP} V_{out} = _____ V_{PP}

14. Using the values recorded in step 13, calculate the differential voltage gain (A_{vd}) of the circuit.

    A_{vd} = _____

15. Modify your circuit as shown in Figure 35.2.

16. Adjust R_7 so that $V_{B(Q1)}$ and $V_{B(Q2)}$ are at 0 Vdc. Then, repeat step 3 of the procedure to balance the amplifier.

17. After balancing the amplifier, measure and record the voltage from the collector of Q_2 to ground.

    $V_{C(Q2)}$ = _____

**Step 12
Waveforms**

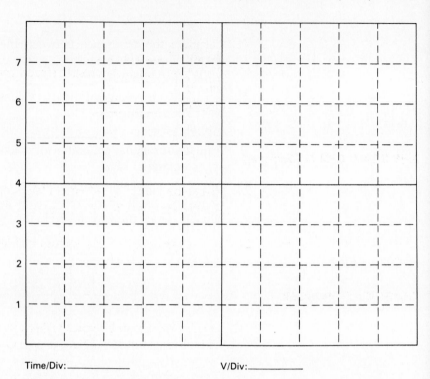

Time/Div:_____   V/Div:_____

18. Adjust R_7 so that $V_{B(Q1)}$ and $V_{B(Q2)}$ are at $+5$ Vdc. Measure and record the voltage from the collector of Q_2 to ground.

$\Delta V_B = $ _____.

19. Using the values from steps 16 and 18, determine the change in V_B.

$V_{C(Q2)} = $ _____

20. Using the values of $V_{C(Q2)}$ from steps 17 and 18, calculate the change in $V_{C(Q2)}$.

$\Delta V_{C(Q2)} = $ _____

21. Using the values obtained in steps 19 and 20, calculate the common mode gain of the amplifier, as follows:

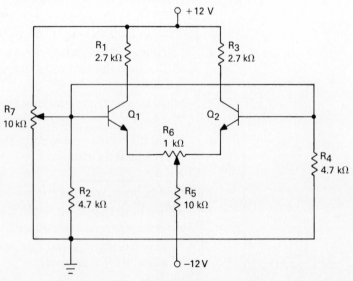

Figure 35.2

$$A_{cm} = \frac{\Delta V_{C(Q2)}}{\Delta V_B}$$

A_{cm} = _____

22. Using the values obtained in steps 14 and 21, calculate the CMRR of the circuit.

    CMRR = _____

THE BRAIN DRAIN (Optional)

23. Prove the following statement to be true:

    A differential amplifier will not work if either of its inputs is not provided with an input current path.

24. In a separate report, discuss your testing procedure and your results.

QUESTIONS/PROBLEMS

1. In terms of circuit operation, discuss the input/output phase relationship observed in step 11 of the procedure.

2. In terms of circuit operation, discuss the input/output phase relationship observed in step 12 of the procedure.

3. In terms of circuit operation, explain the relatively high CMRR of the circuit.

4. Discuss, in your own words, what you observed in this exercise.

Exercise 38

The Voltage Follower

OBJECTIVE

- To demonstrate the operation of the voltage follower.

DISCUSSION

The voltage follower is an op-amp circuit that is most often used as a buffer. The circuit has no phase inversion, unity voltage gain, high input impedance, and low output impedance.

In terms of construction, the voltage follower is one of the simplest of the op-amp circuits. In this exercise, we will take a very brief look at the operation of this circuit.

LAB PREPARATION

Review section 12.5 of *Introductory Electronic Devices and Circuits*.

LAB OVERVIEW

In this exercise, you will:

1. Observe the input/output phase relationship of the voltage follower.
2. Measure the voltage gain of a voltage follower.

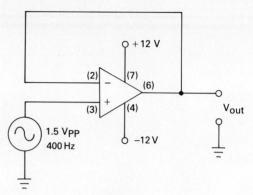

Figure 38.1

MATERIALS

1   Dual-polarity variable dc power supply
1   Variable ac signal source
1   Dual-trace oscilloscope
1   μA741 op-amp (or equivalent)

PROCEDURE

1. Construct the circuit shown in Figure 38.1.
2. Adjust the output of the signal generator for a 1 V_{PP} output at a frequency of 500 Hz.
3. Adjust your oscilloscope to observe the circuit input and output waveforms simultaneously. Draw the input/output waveforms in the space provided below.

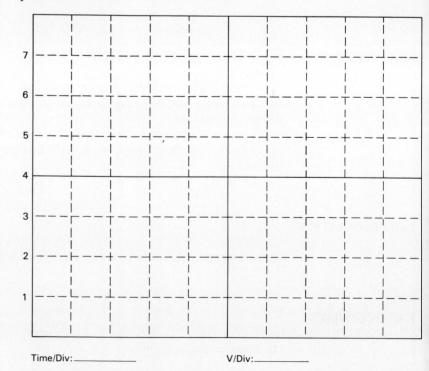

Time/Div:_____   V/Div:_____

4. Measure and record the circuit input and output peak-to-peak voltages.
    V_{in} = _____ V_{PP} V_{out} = _____ V_{PP}

5. Using the values measured in step 4, calculate the voltage gain of the circuit.

$A_{CL} =$ _____

QUESTIONS/PROBLEMS

1. Discuss, in your own words, what you observed in this exercise.

PART VI
Amplifier Frequency Response

Measuring Amplifier Bandwidth

OBJECTIVE

- To demonstrate a practical method for measuring the bandwidth of an amplifier.

DISCUSSION

The *bandwidth* of an amplifier is the range of frequencies over which the gain of the circuit remains relatively constant. The limits of amplifier bandwidth are the *lower cutoff frequency* (f_1) and the *upper cutoff frequency* (f_2). At these frequencies, the voltage and power gain of the amplifier will have dropped by 3 dB. This means that the voltage gain of the amplifier will be approximately equal to $0.5A_{p(mid)}$ and the voltage gain of the amplifier will be approximately equal to $0.707A_{V(mid)}$. Note that the cutoff frequencies of an amplifier are often referred to as the *3 dB points* or the *half-power points*.

In this exercise, you will not perform any involved calculations. Our purpose is to develop a technique for measuring bandwidth. The circuit that you will be using is shown in Figure 39.1. A few points need to be made regarding this circuit.

The circuit shown is different than any you have seen in two respects. First, there is no output coupling capacitor. The output coupling capacitor has been eliminated so that only the *input* coupling capacitor will weigh into the value of f_1. Also, a 470 pF capacitor has been placed in parallel with the amplifier output. This has been done to bring f_2 down to a value that is easily measured. Without this capacitor, the value of f_2 would be in the MHz range. By adding the 470 pF capacitor, the value of f_2 has been decreased to less than 100 kHz. This will be much easier for you to measure using standard test equipment.

LAB PREPARATION

Review section 13.1 of *Introductory Electronic Devices and Circuits*.

LAB OVERVIEW

In this exercise, you will:

1. Measure the lower cutoff frequency of an amplifier.
2. Measure the upper cutoff frequency of the amplifier.
3. Use the measured values of f_1 and f_2 to calculate the bandwidth and center frequency of the amplifier.

MATERIALS

1 Variable dc power supply
1 Variable ac signal generator
1 VOM or DMM
1 Oscilloscope
1 2N3904 npn transistor
4 Resistors: 1 kΩ, 2.2 kΩ, 3.6 kΩ, and 4.7 kΩ
1 10 kΩ potentiometer
3 Capacitors: 470 pF, 1 μF, and 100 μF

PROCEDURE

1. Construct the circuit shown in Figure 39.1.
2. Apply power to the circuit and adjust R_{1b} to provide midpoint bias.
3. Set your ac signal generator for a minimum output amplitude and an operating frequency of approximately 20 kHz.

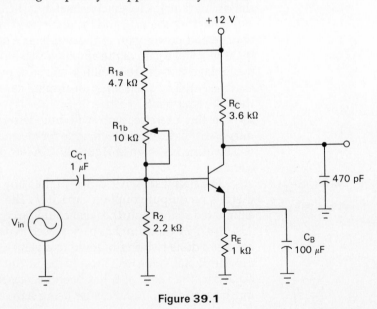

Figure 39.1

4. Connect the oscilloscope to the output of the amplifier and increase the input amplitude until you obtain the maximum undistorted output from the amplifier.

5. Vary the input frequency by several kHz both above and below its present setting to verify that you are operating in midband. If you are, the output amplitude will not change significantly as you vary the input frequency. If you are not in the midband frequency range of the amplifier, adjust f_{in} until you achieve midband operation.

6. Measure and record the peak-to-peak output from the amplifier.

   V_{out} = _____ V_{PP}

7. To measure the value of f_1, decrease the circuit input frequency until the peak-to-peak output voltage equals 70.7% of the value recorded in step 6. Measure and record the frequency at which this occurs.

   f_1 = _____

8. To measure the value of f_2, increase the circuit input frequency until the peak-to-peak output voltage again equals 70.7% of the value recorded in step 6. Measure and record the frequency at which this occurs.

   f_2 = _____

9. Using the values recorded in steps 7 and 8, calculate the following values:

   BW = _____

   f_0 = _____

Exercise 40

Amplifier Bandwidth and Roll-Off Rates

OBJECTIVES

- To demonstrate the frequency response characteristics of a BJT amplifier.
- To demonstrate the frequency response characteristics of an FET amplifier.

DISCUSSION

In exercise 42, you were shown how to measure the bandwidth of an amplifier. In this exercise, you will measure the bandwidth and roll-off rate of a BJT amplifier and an FET amplifier.

LAB PREPARATION

Review sections of 13.3 through 13.5 of *Introductory Electronic Devices and Circuits*.

LAB OVERVIEW

In this exercise, you will:

1. Measure the voltage gain of a BJT amplifier at midband, f_1, $f_1/2$, and $f_1/4$.
2. Measure the bandwidth of the BJT amplifier at f_2, $2f_2$, and $4f_2$.
3. Repeat the steps above for an FET amplifier.
4. Draw the Bode plots for the two amplifiers.

MATERIALS

    1 Variable dc power supply
    1 Variable ac signal generator
    1 VOM or DMM
    1 Oscilloscope
    1 2N3904 npn transistor
    1 2N5485 n-channel JFET

> Note: You will need an X10 oscilloscope probe for all measurements in this exercise.

    6 Resistors: 1 kΩ (2), 3.9 KΩ, 5.1 kΩ, 10 kΩ, and 1 MΩ
    1 100 kΩ potentiometer
    6 Capacitors: 470 pF, 3.3 nF, 1 μF, 22 μF, 100 μF, and 470 μF
    1 Resistor: 3.3 kΩ (Optional)

PROCEDURE

Part I. BJT Amplifier Frequency Response

1. Construct the circuit shown in Figure 40.1.
2. Apply power to the circuit and adjust R_{1b} to provide midpoint bias.
3. Set the signal generator output frequency to approximately 5 kHz. With the amplitude of the signal generator set to minimum, connect it to the amplifier input. Then, slowly increase the signal amplitude until you obtain the maximum undistorted output from the amplifier. Measure and record the following:

   V_{in} = _____ V_{PP} V_{out} = _____ V_{PP}

4. Using the values found in step 3, calculate the voltage gain of the amplifier.

   A_v = _____

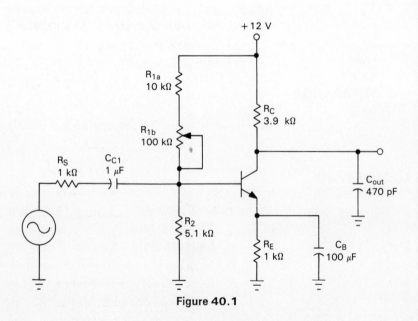

Figure 40.1

5. Decrease the input frequency until f_1 is reached. Measure and record the following:

V_{in} = _____ V_{PP} V_{out} = _____ V_{PP}

f_1 = _____ A_v = _____

6. Decrease the frequency to one-half the value of f_1. Measure and record the following values:

V_{in} = _____ V_{PP}

V_{out} = _____ V_{PP}

A_v = _____

7. Decrease the input frequency to one-fourth the value of f_1. Measure and record the following:

V_{in} = _____ V_{PP}

V_{out} = _____ V_{PP}

A_v = _____

8. Increase the input frequency until f_2 is reached. Measure and record the following:

V_{in} = _____ V_{PP} V_{out} = _____ V_{PP}

f_2 = _____ A_v = _____

9. Increase the input frequency to twice the value of f_2. Measure and record the following:

V_{in} = _____ V_{PP}

V_{out} = _____ V_{PP}

A_v = _____

10. Increase the operating frequency to four times the value of f_2. Measure and record the following:

V_{in} = _____ V_{PP} V_{out} = _____ V_{PP}

A_v = _____

11. Disconnect power from the circuit and measure the value of $(R_{1a} + R_{1b})$.

R_1 = _____

Part II. FET Amplifier Frequency Response

12. Construct the circuit shown in Figure 40.2.

13. Repeat steps 3 through 10 for this circuit. Record your measurements in the appropriate spaces below.

f_1 = _____ f_2 = _____

	V_{in}	V_{out}	A_v
Midband:			
At f_1:			
At $f_1/2$:			
At $f_1/4$:			
At f_2:			
At $2f_2$:			
At $4f_2$:			

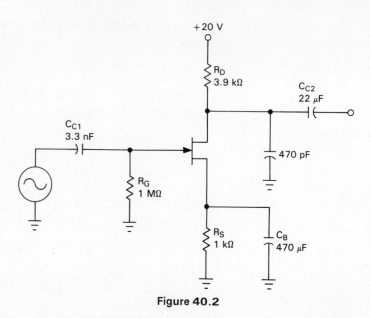

Figure 40.2

THE BRAIN DRAIN (Optional)

14. Calculate the following for the circuit in Figure 40.2:
    a. the theoretical value of f_0.
    b. the actual value of f_0 (using the measured values of f_1 and f_2).
    c. the percent of error between (a) and (b) above.
15. Predict the shift in f_0 that will occur if a 3.3 kΩ load resistance is added to the amplifier. Use the actual value of f_0 for the unloaded circuit as a reference value.
16. Add the load to the circuit and determine the actual shift in f_0.
17. In a separate report, include:
    a. all predicted and measured values.
    b. the percent of error from step 14c.
    c. the percent of error between the predicted and actual shift in f_0.
    d. an explanation of the percents of error found.

QUESTIONS/PROBLEMS

1. The lower cutoff frequency of the circuit in Figure 40.1 is determined by its base circuit. The value of f_{1B} is found as

$$f_{1B} = \frac{1}{2\pi(R_S + R_{in})C_{C1}}$$

    where

$$R_{in} = R_1 \| R_2 \| h_{ie}$$

    Using $h_{ie} = 2$ kΩ, calculate the value of f_{1B} for the amplifier in Figure 40.1. (Note: $h_{ie} = 2$ kΩ is a typical value for a 2N3904.)
    $f_{1B} = $ _____

2. Calculate the percent of error between the calculated value of f_{1B} and the value of f_1 measured in step 5 of the procedure.
    % of error = _____

How would you account for this error?

3. The upper cutoff frequency of the amplifier in Figure 40.1 is determined by its collector circuit. The value of f_{2C} for the circuit is found as

$$f_{2C} = \frac{1}{2\pi R_c C_{out}}$$

Calculate the value of f_{2C} for the circuit in Figure 40.1.

$f_{2C} =$ _____

4. Calculate the percent of error between the calculated value of f_{2C} and the value of f_2 measured in step 8 of the procedure.

% of error = _____

How would you account for this error?

5. On a separate sheet of paper, draw the Bode plot for the BJT amplifier in Figure 40.1. Use your measured frequency and gain values.

6. Repeat problem 5 for the FET amplifier in Figure 40.2.

7. Discuss, in your own words, what you observed in this exercise.

Exercise 41

Op-Amp Slew Rates

OBJECTIVES

- To demonstrate the effect of op-amp slew rate on the frequency response of an inverting amplifier.
- To demonstrate the effect of peak output voltage on the frequency response of an inverting amplifier.

DISCUSSION

The maximum operating frequency of an inverting (or noninverting) amplifier is limited by two factors: the *slew rate* of the op-amp and the *peak output voltage* from the amplifier. The slew rate of the op-amp is the maximum rate at which the output voltage from the device can change without distorting the input waveform.

The relationship between f_{max}, op-amp slew rate, and peak output voltage is given as:

$$f_{max} = \frac{slew\ rate}{2\pi V_{pk}}$$

When the value of f_{max} for a given inverting amplifier is exceeded, *slew rate distortion* occurs. The effect of slew rate distortion on a sinusoidal waveform is shown in Figure 41.1. Note that the input sine wave is distorted into a sawtooth output waveform.

The slew rate distortion shown in Figure 41.1 can be eliminated by:

1. decreasing the input frequency to the circuit.
2. decreasing the input amplitude.
3. using an op-amp with a higher slew rate.

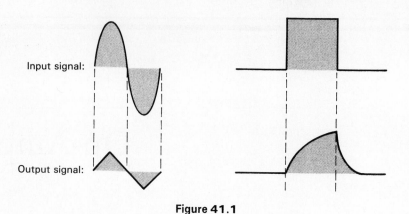

Input signal:

Output signal:

Figure 41.1

LAB PREPARATION

Review section 13.6 of *Introductory Electronic Devices and Circuits.*

LAB OVERVIEW

In this exercise, you will:

1. Measure the slew rate of your op-amp.
2. Predict and measure the maximum operating frequency of an inverting amplifier with a given peak output voltage.
3. Decrease the peak output voltage from the inverting amplifier and observe the effect on the maximum operating frequency of the circuit.

MATERIALS

1   Dual-polarity variable dc power supply
1   Variable ac signal generator
1   Variable square wave generator
1   Dual-trace oscilloscope
1   μA741 op-amp or equivalent
2   Resistors: 1 kΩ and 10 kΩ
1   TL071 op-amp (optional)

PROCEDURE

1. Construct the circuit shown in Figure 41.2.
2. Apply a 100 kHz sine wave to the amplifier input. Adjust the amplitude of the input so that the output transition equals 10 V_{PP}.

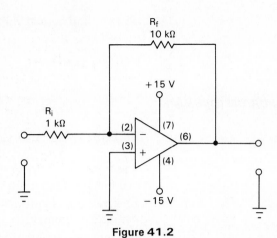

Figure 41.2

3. Adjust your oscilloscope so that you can measure the time required for the output to measure the 10 V output transition.

4. Verify that you are observing the minimum transition time using the following steps:

   a. vary the input amplitude to the circuit while observing the slope of the output transition. If the slope of the output does not vary with input amplitude, you are observing the minimum output transition time.

   c. If the slope of the output signal *does* vary, you need to increase the input frequency until it becomes constant.

   d. Once you have a constant output slope, return the amplitude of the input to the value required to obtain a 10 V_{PP} output from the circuit.

5. Measure and record the time required for the output of the circuit to make the 10 V_{PP} transition.

   Transition time = _____

6. Using the value of 10 V_{PP} and the required transition time, calculate the slew rate of the op-amp in $V/_{\mu s}$.

   Slew rate = _____

7. Using the calculated slew rate of your op-amp and V_{pk} = +10 V, predict the maximum operating frequency of the circuit.

   f_{max} = _____

8. Set your ac signal generator for a 10 V_{pk} output at a frequency of 1 kHz.

9. Apply the ac input signal to the circuit. Increase the operating frequency until you begin to observe slew rate distortion at the circuit output.

10. Measure and record the frequency at which the slew rate distortion begins to occur.

    f_{max} = _____

11. Using the calculated slew rate of your op-amp and V_{pk} = +5 V, predict the maximum operating frequency of the circuit.

    f_{max} = _____

12. Set your ac signal generator for a 5 V_{pk} output at a frequency of 1 kHz.

13. Apply the ac signal to the circuit. Increase the operating frequency until you begin to observe slew rate distortion at the circuit output.

14. Measure and record the frequency at which the slew rate distortion begins to occur.

    f_{max} = _____

THE BRAIN DRAIN (Optional)

15. The TL071 has a rated slew rate of 13 V/μs. Assuming that a TL071 is being used in a noninverting amplifier with values of R_f = 10 kΩ, R_i = 1 kΩ, and V_{in} = 2 V_{PP}, predict the maximum operating frequency of the circuit. Assume that the circuit has a sinusoidal input.

16. Construct the circuit described in step 15 and measure its maximum operating frequency.

17. In a separate report, include your circuit schematic, predicted and measured values of f_{max}, the percent of error between the two values, and an explanation of that error.

QUESTIONS/PROBLEMS

1. The μA741 has a rated slew rate of 0.5 V/μs. What is the percent of error between this value and the one measured in steps 2 through 6 of the procedure?

   % of error = _____

   How would you account for this error?

2. What is the percent of error between the values of f_{max} found in steps 7 and 10 of the procedure?

   % of error = _____

   How would you account for this error?

3. Compare the results of steps 10 and 14 of the procedure. Based on your results, what is the relationship between maximum operating frequency and peak output voltage?

4. Discuss, in your own words, what you observed in this exercise.

PART VII

Negative Feedback Amplifiers

The Effects of Feedback on Amplifier Bandwidth

OBJECTIVE

- To observe the effects of negative voltage and current feedback on the bandwidth of an op-amp.

DISCUSSION

In Exercises 36 and 37, you were shown the effects of negative voltage feedback on the gain characteristics of an op-amp. In this exercise, you will be shown the effects of noninverting voltage feedback and noninverting current feedback on the bandwidth of an op-amp.

LAB PREPARATION

Review sections 14.3 and 14.5 of *Introductory Electronic Devices and Circuits*.

LAB OVERVIEW

In this exercise, you will:

1. Calculate the feedback factor for a noninverting voltage feedback amplifier and predict the bandwidth of the circuit.
2. Construct the circuit and measure its bandwidth.
3. Repeat steps 1 and 2 for a noninverting current feedback amplifier.

MATERIALS

1   Dual-polarity variable dc power supply
1   Variable ac signal generator
1   Oscilloscope
1   μA741 op-amp or equivalent (along with its specification sheet)
4   Resistors: 1 kΩ, 9.1 kΩ, 10 kΩ, and 22 kΩ
1   TL071 op-amp (optional)

PROCEDURE

1. Obtain the following ratings from the spec sheet of your op-amp:
    A_{OL} = _____ f_{unity} = _____
    Note that the f_{unity} rating may be listed as *bandwidth* on your spec sheet.
2. Construct the circuit in Figure 42.1a.
3. Using the appropriate equations from section 14.3 of the text, calculate
    the following values for the amplifier:

    α_v = _____
    A_{CL} = _____
    BW = _____
4. Apply power to the circuit, then, apply a 1 V_{PP} input to the circuit. Set
    the input frequency to approximately 1 kHz.
5. Measure and record the gain of the amplifier.

    A_{CL} = _____
6. Measure and record the value of f_2 for the amplifier. Remember, since
    the amplifier is a dc amplifier, the overall circuit bandwidth equals f_2.

    f_2 = _____

    > Note: Slew rate distortion may
    > occur well before f_2 is reached.

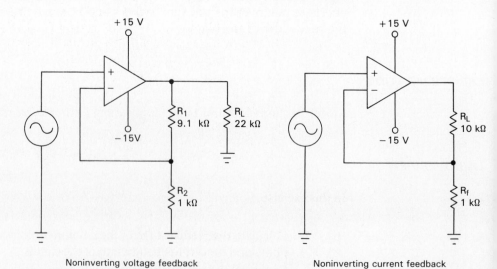

Noninverting voltage feedback   Noninverting current feedback
(a)   (b)

Figure 42.1

Part II. Noninverting Current Feedback

7. Construct the circuit shown in Figure 42.1b.

8. Using the appropriate equations in section 14.5 of the text and the op-amp ratings listed in step 1, calculate the following values for the circuit.

    α_i = _____

    A_{CL} = _____

    BW = _____

9. Apply power to the circuit. Then, apply a 1 V_{PP}, 800 Hz input signal to the circuit.

10. Measure and record the value of A_{CL} for the circuit.

    A_{CL} = _____

11. Measure and record the value of f_2 (and thus, bandwidth) for the circuit.

    BW = _____

THE BRAIN DRAIN (Optional)

12. The TL071 has a f_{unity} rating of 3 MHz. Using the TL071 in a noninverting voltage feedback configuration, prove that the following relationship holds true:

$$f_{unity} = A_{CL}f_2$$

Use R_i = 1 kΩ and your other resistors as a series of R_f values. Do not place a load on the circuit.

13. In a separate report, include your predicted and measured values, the percents of error between your predicted and measured values, explanations of those errors, and a comparison of the f_{unity} obtained by multiplying the measured values of A_{CL} and f_2.

QUESTIONS/PROBLEMS

1. For the noninverting voltage feedback amplifier, calculate the percents of error between your predicted and measured values of A_{CL} and BW.

    % of error in A_{CL} = _____

    % of error in BW = _____

2. Repeat question 1 for the noninverting current feedback amplifier.

    % of error in A_{CL} = _____

    % of error in BW = _____

3. How would you account for the percents of error obtained in question 1?

4. How would you account for the percents of error obtained in question 2?

5. For a given voltage or current feedback amplifier, the value for A_{CL} can be approximated as:

$$A_{CL} = \frac{1}{\alpha_v} \qquad\qquad \text{or} \qquad\qquad A_{CL} = \frac{1}{\alpha_i}$$

depending on the type of amplifier being analyzed. Are these approximations valid for the circuits in this exercise? Explain your answer using calculations based on the data in this exercise.

6. Discuss, in your own words, what you observed in this exercise.

PART VIII
Tuned Amplifiers

Exercise 43

Low-Pass Active Filters

OBJECTIVES

- To demonstrate the operation of a single-pole low pass-active filter.
- To demonstrate the operation of a two-pole low-pass active filter.

DISCUSSION

Active filters are tuned op-amp circuits that contain one or more *poles*. A *pole* is nothing more than a single RC circuit.

The roll-off rate for an active filter depends on the number of poles it contains. For example, a Butterworth active filter has a roll-off rate of *6 dB per octave* (or *20 dB per decade*) per pole. Thus, a two-pole Butterworth filter has a roll-off rate of approximately 12 dB per octave, or 40 dB per decade.

In this exercise, you will be observing the operation of a one-pole low-pass filter and a two-pole low pass filter. Then, in the exercises that follow, you will observe the operation of some high-pass, bandpass, and notch filters.

LAB PREPARATION

Review section 15.2 of *Introductory Electronic Devices and Circuits*.

LAB OVERVIEW

In this exercise, you will:

1. Analyze the gain and frequency response characteristics of a one-pole low-pass filter.
2. Analyze the gain and frequency response characteristics of a two-pole low-pass filter.

MATERIALS

1   Dual-polarity variable dc power supply
1   Variable ac signal generator
1   Oscilloscope
1   μA741 op-amp or equivalent
3   Resistors: 15 kΩ (2), 30 kΩ
2   Capacitors: 0.01 μF, 0.022 μF
1   μA741 op-amp or equivalent (additional-optional)
2   15 kΩ resistors (additional-optional)

PROCEDURE

1. Construct the circuit shown in Figure 43.1. Initially, the ac signal generator should be set to its minimum output amplitude.
2. Using equation 15.5 in the text, predict the upper cutoff frequency of the circuit.

   f_2 = _____
3. Apply a 1 V_{PP} input to the circuit at one-fourth the frequency calculated in step 2.
4. Measure and record the following values:

   f_{in} = _____

   A_{CL} = _____
5. Convert the A_{CL} value in step 4 to dB form.

   A_{CL} = _____ dB

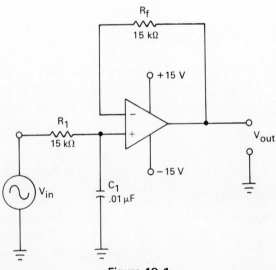

Figure 43.1

6. Measure and record the value of f_2 for the filter.

    $f_2 =$ _____

7. Adjust the input frequency to $2f_2$. Measure and record the voltage gain of the circuit.

    $A_{CL} =$ _____

8. Convert the A_{CL} value in step 7 to dB form.

    $A_{CL} =$ _____dB

9. Adjust the input frequency to $4f_2$. Measure and record the voltage gain of the circuit.

    $A_{CL} =$ _____

10. Convert the A_{CL} value in step 9 to dB form.

    $A_{CL} =$ _____dB

Part II. The Two-Pole Low-Pass Filter

11. Construct the circuit shown in Figure 43.2.

12. Using equation (15.6) in the text, predict the upper cutoff frequency of the circuit.

    $f_2 =$ _____

13. Set the input frequency to one-fourth the value obtained in step 12. Again, set the amplitude of the input signal to 1 V_{PP}.

14. Measure and record the following values:

    $f_{in} =$ _____

    $A_{CL} =$ _____

15. Convert the value of A_{CL} in step 14 to dB form.

    $A_{CL} =$ _____dB

16. Measure and record the value of f_2 for the filter.

    $f_2 =$ _____

17. Adjust the input frequency to $2f_2$. Measure and record the voltage gain of the filter.

    $A_{CL} =$ _____

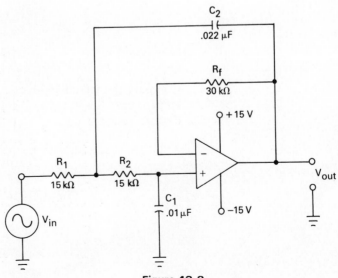

Figure 43.2

18. Convert the value in step 17 to dB form.

    A_{CL} = _____dB

19. Adjust the input frequency to $4f_2$. Measure and record the voltage gain of the filter.

    A_{CL} = _____

20. Convert the value in step 19 to dB form.

    A_{CL} = _____dB

THE BRAIN DRAIN (Optional)

21. Cascade the circuits in Figures 43.1 and 43.2 to form a three-pole low-pass filter.

22. Perform a complete frequency versus gain analysis on the three-pole filter.

23. In a separate report, include the following:
    a. the predicted and measured cutoff frequency values.
    b. your gain and frequency measurements.
    c. a Bode plot for the filter.
    d. the percent of error between your predicted and measured values of f_2, along with an explanation of the error.

QUESTIONS/PROBLEMS

1. Calculate the percent of error between your predicted and measured values of f_2 for the one-pole filter.

   % of error = _____

   How would you account for this error?

2. Compare the values in steps 5, 8, and 10 of the procedure. Do these values verify the stated 6 dB per octave roll-off rate for a one-pole filter? Explain your answer.

3. Calculate the percent of error between your predicted and measured values of f_2 for the two-pole filter.

    % of error = _____

    How would you account for this error?

4. Compare the values in steps 15, 18, and 20. Do these values verify the stated roll-off rate of 12 dB per octave for a two-pole filter? Explain your answer.

5. Discuss, in your own words, what you observed in this exercise.

Exercise 44

High-Pass Active Filters

OBJECTIVES

- To demonstrate the operation of the one-pole high-pass filter.
- To demonstrate the operation of the two-pole high-pass filter.

DISCUSSION

High-pass active filters are designed to pass all frequencies *above* a predetermined lower cutoff frequency (f_1). The stated roll-off rates for high-pass filters are the same as those for the low pass filter: *6 dB per octave* (or *20 dB per decade*) per pole. Note that the same equations are used to determine the value of f_1 for a high-pass filter and the value of f_2 for the low-pass filter.

LAB PREPARATION

Review section 15.2 of *Introductory Electronic Devices and Circuits*.

LAB OVERVIEW

In this exercise, you will:

1. Analyze the gain and frequency response characteristics of a one-pole high-pass filter.
2. Analyze the gain and frequency response characteristics of a two-pole active filter.

MATERIALS

    1 Dual-polarity variable dc power supply
    1 Variable ac signal generator
    1 Oscilloscope
    1 μA741 op-amp or equivalent
    3 Resistors: 10 kΩ (2), 22 kΩ
    2 0.01 μF capacitors
    1 30 kΩ resistor (additional-optional)

PROCEDURE

1. Construct the circuit shown in Figure 44.1. Initially, the ac signal generator should be set to its minimum output amplitude.
2. Using equation 15.5 in the text, predict the lower cutoff frequency of the circuit.

    f_1 = _____
3. Apply a 1 V_{PP} input to the circuit at four times the frequency calculated in step 2.
4. Measure and record the following values:

    f_{in} = _____

    A_{CL} = _____
5. Convert the A_{CL} value in step 4 to dB form.

    A_{CL} = _____dB
6. Measure and record the value of f_1 for the filter.

    f_1 = _____
7. Adjust the input frequency to $f_1/2$. Measure and record the voltage gain of the filter at this frequency.

    A_{CL} = _____
8. Convert the value in step 7 to dB form.

    A_{CL} = _____dB

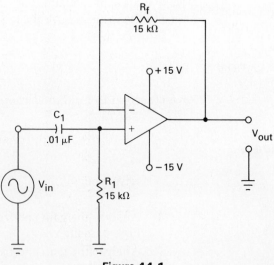

Figure 44.1

9. Adjust the input frequency to $f_1/4$. Measure and record the voltage gain of the filter at this frequency.

   A_{CL} = _____

10. Convert the value in step 9 to dB form.

    A_{CL} = _____dB

Part II. The Two-Pole High-Pass Filter

11. Construct the circuit shown in Figure 44.2.

12. Using equation (15.6) in the text, predict the lower cutoff frequency of the circuit.

    f_1 = _____

13. Set the input frequency to four times the value obtained in step 12. Again, set the amplitude of the input signal to 1 V_{PP}.

14. Measure and record the following values:

    f_{in} = _____

    A_{CL} = _____

15. Convert the value of A_{CL} in step 14 to dB form.

    A_{CL} = _____dB

16. Measure and record the value of f_1 for the circuit.

    f_1 = _____

17. Adjust the input frequency to $f_1/2$. Measure and record the voltage gain of the filter.

    A_{CL} = _____

18. Convert the value in step 17 to dB form.

    A_{CL} = _____dB

19. Adjust the input frequency to $4f_2$. Measure and record the voltage gain of the filter.

    A_{CL} = _____

20. Convert the value in step 19 to dB form.

    A_{CL} = _____dB

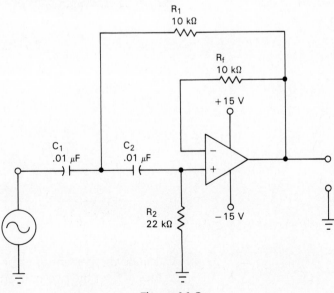

Figure 44.2

THE BRAIN DRAIN (Optional)

21. Using Figure 15.19 in the text as a model, convert your two-pole high-pass filter to a filter with a gain of $A_{CL} = 4$.
22. Using input frequencies of $4f_1$, $3f_1$, $2f_1$, f_1, $f_1/2$, and $f_1/4$, measure the dB voltage gain of the amplifier.
23. In a separate report, compare the Bode plot of this circuit to that of the two-pole filter in Figure 44.2. Include a discussion of the differences between the two Bode plots.

QUESTIONS/PROBLEMS

1. Calculate the percent of error between your predicted and measured values of f_1 for the one-pole filter.

   % of error = _____

   How would you account for this error?

2. Compare the values in steps 5, 8, and 10 of the procedure. Do these values verify the stated 6 dB per octave roll-off rate for a one-pole filter?

3. Calculate the percent of error between your predicted and measured values of f_1 for the two-pole filter.

   % of error = _____

How would you account for this error?

4. Discuss, in your own words, what you observed in this exercise.

Exercise 45

The Multiple-Feedback Bandpass Filter

OBJECTIVE

- To demonstrate the frequency response characteristics of the multiple-feedback bandpass filter.

DISCUSSION

The multiple-feedback bandpass filter is designed using a single op-amp and multiple RC circuits. While this type of filter is easier to construct than its multi-stage counterpart, it is subject to several limitations. First, the multiple-feedback circuit is generally restricted to values of Q that are less than 50. Second, the gain of a multiple-feedback filter cannot be independently controlled. For example, the value of A_{CL} for the circuit in Figure 45.1 can be approximated as

$$A_{CL} = \frac{R_f}{2R_1}$$

Since the values of R_f and R_1 both weigh into the f_0 equation for the circuit, you cannot change A_{CL} without changing the frequency response characteristics of the circuit.

LAB PREPARATION

Review section 15.3 of *Introductory Electronic Devices and Circuits*.

LAB OVERVIEW

In this exercise, you will:

1. Predict the values of f_0, Q, BW, f_1, and f_2 for a multiple-feedback bandpass filter.
2. Measure the values of f_1 and f_2 for the circuit.
3. Using the measured values of f_1 and f_2, determine the actual values of f_0 and Q for the circuit.

MATERIALS

1   Dual-polarity variable dc power supply
1   Variable ac signal generator
1   Oscilloscope
1   μA741 op-amp or equivalent
3   Resistors: 1 kΩ, 10 kΩ, and 180 kΩ
2   0.001 μF capacitors

> Note: See Figure 15.58 (page 735 of the text) to determine the parts needed for the optional brain drain problem.

PROCEDURE

1. Construct the circuit shown in Figure 45.1.
2. Using equation (15.6) in the text, predict the value of f_0 for the circuit.
   f_0 = _____
3. Using equation (15.7) in the text, predict the value of Q for the circuit.
   Q = _____
4. Using your values of f_0 and Q, predict the following values for the filter.
   BW = _____
   f_1 = _____
   f_2 = _____
5. Apply power to the circuit. Set your ac signal generator amplitude to its minimum value. Set the frequency of the signal generator to approximately 12 kHz.

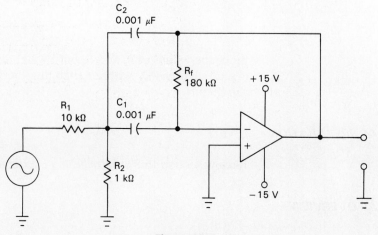

Figure 45.1

6. While viewing the output of the circuit, increase the circuit input amplitude until you obtain an output of approximately 4 V_{PP}.

7. Vary the input frequency until you obtain the maximum output amplitude from the filter. Then, readjust the input amplitude to obtain a 4 V_{PP} output from the filter.

8. Measure and record the following values:

   f_1 = _____ f_2 = _____

9. Using the values obtained in step 8, calculate the actual center frequency of the circuit.

   f_0 = _____

10. Using the values obtained in step 8, calculate the actual bandwidth of the filter.

   BW = _____

11. Using the values of Q and BW from steps 9 and 10, calculate the actual value of Q for the filter.

   Q = _____

THE BRAIN DRAIN (Optional)

12. Figure 15.58 (page 735 of the text) is a multi-stage notch filter. While the circuit works well on paper, there is a flaw in its design. Construct the circuit and, through testing, determine the flaw in the circuit design. Here are a few hints:

    1. The flaw doesn't prevent the circuit from operating. It just doesn't operate very well.
    2. The flaw cannot be isolated to any single component.

13. In a separate report, include your testing procedure, the flaw in the design, and a suggestion for correcting it.

QUESTIONS/PROBLEMS

1. Calculate the percent of error between your predicted and measured values of each of the following:

   % of error in f_0 = _____

   % of error in f_1 = _____

   % of error in f_2 = _____

   % of error in BW = _____

   % of error in Q = _____

2. How would you account for each of the percents of error obtained in question 1?

3. Discuss, in your own words, what you observed in this exercise.

The Discrete Tuned Amplifier

OBJECTIVE

- To demonstrate the frequency response characteristics of a tuned BJT amplifier.

DISCUSSION

A tuned BJT amplifier is one that is designed to have a specific bandwidth that is centered around a specific frequency. The values of f_0 and bandwidth for this type of amplifier are determined by the inductor and capacitor that are used in the collector circuit of the BJT.

When a load is placed on a tuned BJT amplifier, the bandwidth characteristics of the amplifier changes. This is due to the change in the circuit Q caused by the presence of the load.

LAB PREPARATION

Review sections 15.5 and 15.6 of *Introductory Electronic Devices and Circuits*.

LAB OVERVIEW

In this exercise, you will:

1. Predict the center frequency (f_0) of a tuned BJT amplifier.
2. Construct the circuit and measure its values of f_1 and f_2.

3. Use the values of f_1 and f_2 to determine the circuit f_0, BW, and Q.
4. Remove the amplifier load and repeat steps 1 through 3.

MATERIALS

1   Variable dc power supply
1   Variable ac signal generator
1   Oscilloscope
1   VOM or DMM
1   2N3904 npn transistor

> Note: You will need an X10 oscilloscope probe to reduce circuit loading.

4   Resistors: 1 kΩ, 10 kΩ, 12 kΩ, 22 kΩ
4   Capacitors: 0.002 μF, 10 μF (2), 100 μF
1   1 mH inductor
1   100 kΩ potentiometer

PROCEDURE

1. Construct the circuit shown in Figure 46.1.
2. Apply power to the circuit. Adjust R_{1b} so that V_E is approximately 2 V.
3. Measure and record the following:
    V_E = _____ V_{CE} = _____
4. Using the rated values of L_1 and C_1, calculate the center frequency (resonant frequency) of the circuit.
    f_0 = _____
5. Set your ac signal generator to the value found in step 4. Adjust the amplitude of the amplifier input signal so that you obtain the maximum *undistorted* output from the circuit.
6. Vary the amplifier input *frequency* slightly in both directions to obtain the maximum possible peak-to-peak output from the amplifier. If the

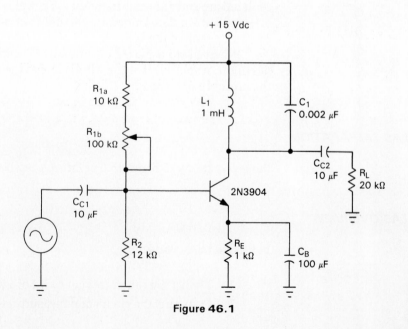

Figure 46.1

maximum output is distorted, adjust the input *amplitude* to eliminate the distortion.

7. Measure and record the following values:

f_1 = _____ f_2 = _____

8. Using the values obtained in step 7, calculate the actual value of each of the following.

f_0 = _____

BW = _____

Q = _____

9. Remove the 20 kΩ load from the amplifier.

10. Adjust the amplifier input/output as described in step 6.

11. Measure and record the following values:

f_1 = _____ f_2 = _____

12. Using the values obtained in step 11, calculate the actual value of each of the following:

f_0 = _____

BW = _____

Q = _____

THE BRAIN DRAIN (Optional)

13. Measure the value of your 20 kΩ resistor.

14. Using the measured value of R_L and the values obtained in step 8, calculate the value of R_w for your inductor (L_1).

15. Using the values obtained in step 12, recalculate the value of R_w for your inductor.

16. Using your DMM (or VOM), measure the value of R_w.

17. In a separate report, include the two sets of R_w calculations, the measured value of R_w, and an explanation of the differences between the three values.

QUESTIONS/PROBLEMS

1. What is the percent of error between the predicted and actual values of the loaded tuned amplifier?

% of error = _____

How would you account for this error?

2. Compare the values obtained in steps 8 and 12 of the procedure. What happens to each of the values when the load is removed?

3. Explain each of the changes described in question 2.

4. Discuss, in your own words, what you observed in this exercise.

Exercise 47

The Class C Amplifier

OBJECTIVE

- To demonstrate the dc and ac operating characteristics of the class C amplifier.

DISCUSSION

The class C amplifier is a circuit that produces an output signal while the transistor conducts for less than 180° of the ac input cycle. Since the transistor is only in conduction for a small portion of the ac input cycle, the efficiency rating of this type of amplifier is extremely high; up to 99%.

Since the class C amplifier uses a resonant LC circuit as a collector circuit, the amplifier is a tuned circuit. The LC circuit is tuned to the circuit input frequency or some even-order harmonic of the input frequency, as you will see in this exercise.

LAB PREPARATION

Review section 15.7 of *Introductory Electronic Devices and Circuits*.

LAB OVERVIEW

In this exercise, you will:

1. Predict and measure the values of V_B, V_C, and V_E for a class C amplifier.

2. Predict and measure the value of f_0 for the circuit.
3. Compare the operation of the circuit at $f_{in} = f_0$ with that of the circuit at $f_{in} = f_0/2$.

MATERIALS

1   Dual-polarity variable dc power supply
1   Variable ac signal generator
1   VOM or DMM
1   Dual-trace oscilloscope
1   2N3904 npn transistor
2   Resistors: 1.5 kΩ, and 22 kΩ
3   Capacitors: 0.002 μF, 10 μF (2)
1   1 mH inductor

PROCEDURE

1. Construct the circuit shown in Figure 47.1.
2. Predict the following values for the circuit:
    V_B = _____
    V_C = _____
    V_E = _____
3. Apply power to the circuit and measure the following values:
    V_B = _____
    V_C = _____
    V_E = _____

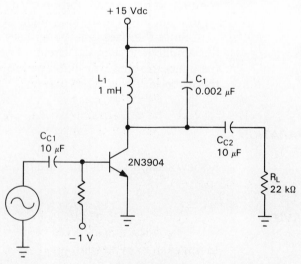

Figure 47.1

4. Predict the value of f_0 for the amplifier.

   $f_0 =$ _____

5. Set the amplitude of your ac signal generator to minimum at a frequency that is approximately equal to the value calculated in step 4.

6. Adjust the amplitude of the amplifier input signal so that you obtain the maximum undistorted output from the circuit.

7. Vary the amplifier input frequency slightly in both directions to obtain the maximum possible peak-to-peak output from the amplifier. If the maximum output is distorted, adjust the input amplitude to eliminate the distortion.

8. Measure and record the actual value of f_0 for the circuit.

   $f_0 =$ _____

9. In the space provided, draw the input and output waveforms. Indicate the positive and negative peak values for both waveforms.

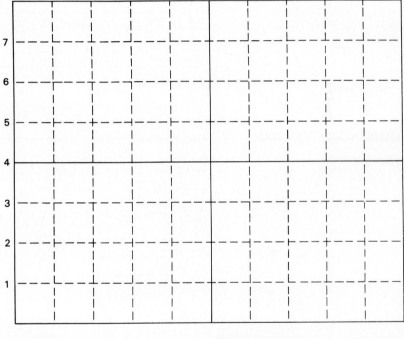

Time/Div: _____   V/Div: _____

10. Decrease the input frequency to approximately one-half the value recorded in step 8. Readjust the circuit as described in steps 6 and 7 of the procedure.

11. In the space provided on the following page, draw the input and output waveforms.

THE BRAIN DRAIN (Optional)

12. Observe the circuit input/output relationship at input frequencies of $f_0/4$, $f_0/6$, and $f_0/8$.

13. In a separate report, plot a curve of input frequency versus amplifier gain, and give an explanation of your results.

**Step 11
Waveforms**

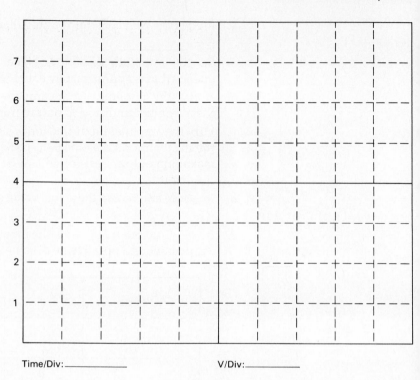

Time/Div:_____   V/Div:_____

QUESTIONS/PROBLEMS

1. What is the percent of error between your predicted and measured values of f_0?

    % of error = _____

    How would you account for this error?

2. Compare the output waveforms drawn in steps 9 and 11 of the procedure. How do they compare in terms of frequency and peak-to-peak value?

3. Discuss, in your own words, what you observed in this exercise.

PART IX
Additional Op-Amp Circuits

Exercise 48

Comparators

- To demonstrate the basic operation of the op-amp comparator.

DISCUSSION

A comparator is a circuit that is used to compare two input voltages and provide a dc output that indicates which of the two inputs is greater. In most cases, the comparator is used to compare one changing input voltage to a set dc reference voltage. For example, a comparator may be used to indicate whether a sine wave is greater than or less than +5 Vdc (or some other dc voltage).

The dc outputs from a comparator are usually referred to as *high* and *low*. Whether the output is high ($+V - 1$ V) or low ($-V + 1$ V) depends on which of the two input voltages is greater.

LAB PREPARATION

Review section 16.1 of *Introductory Electronic Devices and Circuits*.

LAB OVERVIEW

In this exercise, you will:

1. Observe the operation of a comparator with a positive dc reference voltage.

2. Observe the operation of a comparator with a negative dc reference voltage.

MATERIALS

1 Dual-polarity variable dc power supply
1 Variable ac signal generator
1 Dual-trace oscilloscope
1 μA741 op-amp (or equivalent)
2 Resistors: 1 kΩ, 10 kΩ
1 100 kΩ potentiometer (optional)

PROCEDURE

1. Construct the circuit shown in Figure 48.1.
2. Apply a 5 V_{PP}, 1 kHz input signal to the comparator.
3. In this step, you're going to set up your oscilloscope so that both traces are ground referenced to the center of the CRT. This will enable you to use your oscilloscope to measure the value of input voltage that causes the op-amp output to change states. Perform the following procedure:
    a. Set both AC/GND/DC switches to the ground (GND) position.
    b. Move both traces to the center of the CRT.
    c. Set both switches to the DC position.
4. *Without changing the positions of the oscilloscope traces*, observe the input and output waveforms for the circuit. Make sure that the channels of your oscilloscope are set to the same VOLTS/DIV setting. Draw the two waveforms in the space provided on the following page.
5. Locate the point at which the output waveform crosses the input sine wave. The instantaneous value of V_{in} at this intersection is the reference voltage (V_{ref}) for the circuit. Measure and record this voltage.

    $V_{ref} =$ _____

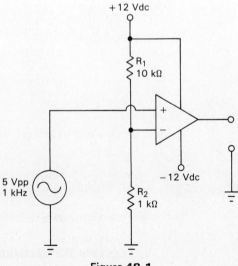

Figure 48.1

**Step 4
Waveforms**

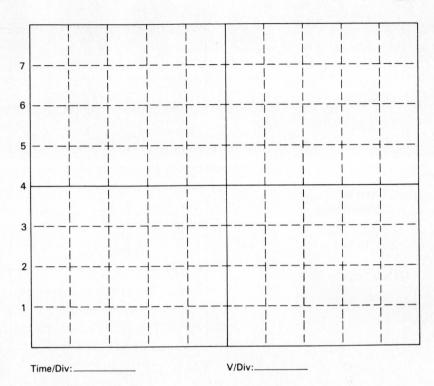

Time/Div:_____   V/Div:_____

6. Describe what happens to the comparator output voltage when the input waveform makes a positive-going transition past V_{ref}.

7. Describe what happens to the comparator output voltage when the input waveform makes a negative-going transition past V_{ref}.

8. Modify your circuit as shown in Figure 48.2.

9. With your traces aligned as described in step 3 of the procedure, observe the input and output waveforms for the circuit. Draw the two waveforms in the space provided on the following page.

Step 9
Waveforms

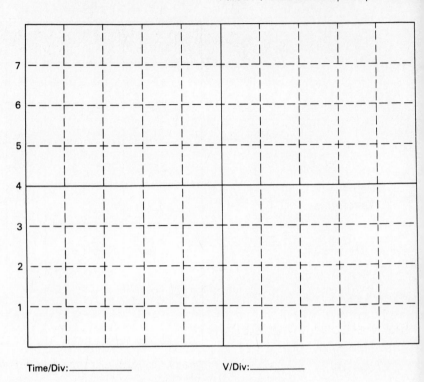

Time/Div:_____   V/Div:_____

10. Measure the value of V_{ref} for the circuit.

$V_{ref} =$ _____

11. Describe what happens to the comparator output voltage when the input waveform makes a positive-going transition past V_{ref}.

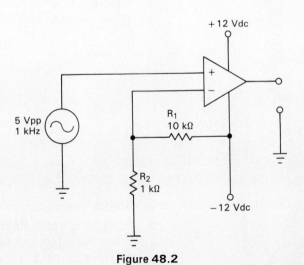

Figure 48.2

12. Describe what happens to the comparator output voltage when the input waveform makes a negative-going transition past V_{ref}.

THE BRAIN DRAIN (Optional)

13. Design an inverting comparator with a reference voltage that can be set to any value between ±10 Vdc.
14. Observe and draw the input/output waveforms for the circuit with V_{ref} values of − 10 Vdc, − 5 Vdc, 5 Vdc, and 10 Vdc.
15. In a separate report, include your circuit schematics and the input/output waveforms for the V_{ref} values listed.

QUESTIONS/PROBLEMS

1. Calculate the value of V_{ref} in Figure 48.1 using the voltage-divider equation.

    V_{ref} = _____

2. Calculate the percent of error between the value calculated in problem 1 and the value measured in step 5 of the procedure.

    % of error = _____

    How would you account for this error?

3. Based on your observations in steps 11 and 12 of the procedure, is the comparator in Figure 48.2 an inverting or noninverting circuit? Explain your answer.

4. Discuss, in your own words, what you observed in this exercise.

Integrators and Differentiators

OBJECTIVES

- To demonstrate the operation of the integrator.
- To demonstrate the operation of the differentiator.

DISCUSSION

The integrator is a circuit that can be used to convert a square wave into a triangular wave. The circuit contains an op-amp with an input resistor and a feedback capacitor. The op-amp provides a constant-current path for the capacitor, allowing it to charge and discharge at a linear rate.

The differentiator is a circuit that contains an op-amp with an input capacitor and a feedback resistor. This circuit can be used to convert a triangular wave into a square wave, as you will see in this exercise.

LAB PREPARATION

Review section 16.2 of *Introductory Electronic Devices and Circuits*.

LAB OVERVIEW

In this exercise, you will:

1. Construct an integrator and observe its effect on a square wave input.

2. Construct a differentiator and observe its effect on a triangular wave input.

MATERIALS

1 Dual-polarity variable dc power supply
1 Variable function generator
1 Dual-trace oscilloscope
1 μA741 op-amp (or equivalent)
2 Resistors: 10 kΩ, 1 kΩ Br, BL, or ε. Jr bL red
2 Capacitors: 0.001 μF, 0.1 μF

PROCEDURE

1. Construct the circuit shown in Figure 49.1.
2. Connect your oscilloscope to observe the input and output waveforms simultaneously. Draw these waveforms in the space provided.

Time/Div: _.1mS_   V/Div: _2 0 v_

3. Set the oscilloscope channel that is connected to the circuit output to dc coupling.
4. While observing the output waveform, remove R_f from the circuit. Describe what happens when the feedback resistor is removed.

0.0133

1000 =

$T \cdot C = RC$

1.2M

Br, Red)

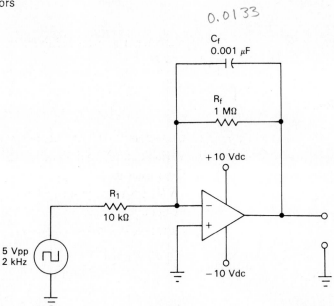

C_f
0.001 μF

R_f
1 MΩ

+10 Vdc

R_1
10 kΩ

5 Vpp
2 kHz

−10 Vdc

Figure 49.1

5. Construct the circuit shown in Figure 49.2.

6. Connect your oscilloscope to observe the input and output waveforms simultaneously. Draw these waveforms in the space provided on the following page.

QUESTIONS/PROBLEMS

1. Explain, in terms of circuit operation, the input/output relationship observed in step 2.

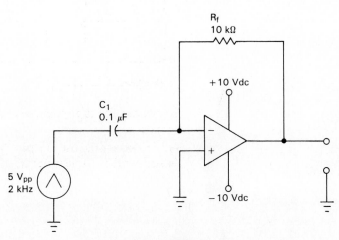

R_f
10 kΩ

+10 Vdc

C_1
0.1 μF

5 Vpp
2 kHz

−10 Vdc

Figure 49.2

**Step 6
Waveforms**

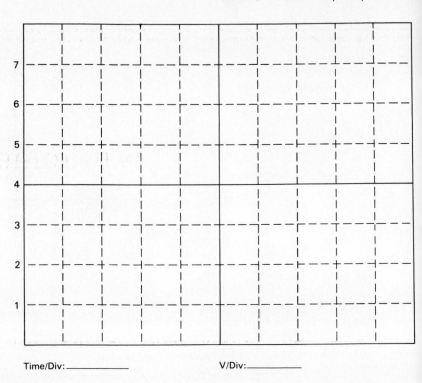

Time/Div:_____   V/Div:_____

2. Explain, in terms of circuit operation, the input/output relationship observed in step 6.

PART X

Oscillators

The Wien-Bridge Oscillator

OBJECTIVE

- To demonstrate the operation of the Wien-bridge oscillator.

DISCUSSION

The Wien-bridge oscillator uses two RC circuits, one series and one parallel, to set the operating frequency of the circuit. The basic Wien-bridge oscillator is shown in Figure 51.1.

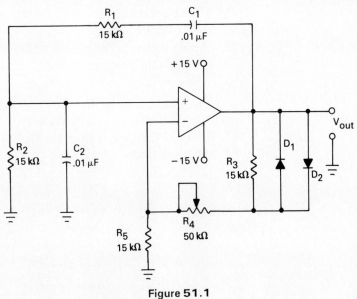

Figure 51.1

As you can see, the series RC circuit has the same component values as the parallel RC circuit. These circuits form a bandpass filter that determines the operating frequency of the oscillator.

The negative feedback circuit is used to control the gain of the op-amp. As you will see, we can vary the output peak voltages by varying the setting of R_4.

LAB PREPARATION

Review sections 17.1 and 17.3 of *Introductory Electronic Devices and Circuits*.

LAB OVERVIEW

In this exercise, you will:

1. Observe the output from a Wien-bridge oscillator.
2. Observe the effects of several component variations on the output of the oscillator.

MATERIALS

1 Dual-polarity variable dc power supply
1 Oscilloscope
1 µA741 op-amp (or equivalent)
1 1N4148 small-signal diodes
4 15 kΩ resistors.
1 50 kΩ potentiometer
2 0.01 µF capacitors

PROCEDURE

1. Construct the circuit shown in Figure 51.1. R_4 should initially be set to approximately 25 kΩ.
2. Apply power to the circuit. Observe the output waveform and draw it neatly in the space provided on the following page.
3. Measure the output frequency from the oscillator.
   $f_{out} = $ _____
4. Vary the setting of R_4 and note the effects on the output waveform.
5. Remove the diodes from the feedback circuit. Note the effects on the circuit output.

**Step 2
Waveform**

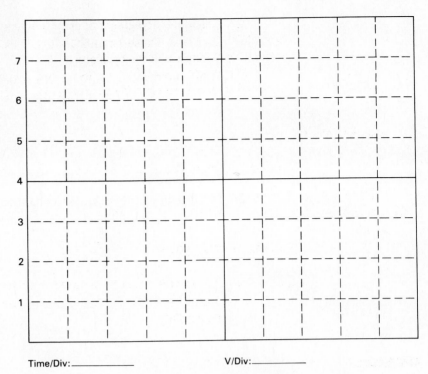

Time/Div:_____   V/Div:_____

6. With the diodes removed, vary the R_4 setting and note any effects on the circuit output.

THE BRAIN DRAIN (Optional)

7. In steps 5 and 6, you saw the effects of removing (opening) D_1 and D_2 in the circuit. Derive a list of the other faults that could occur in the circuit and simulate each of those faults, noting the effects of the circuit output.

8. In a separate report, include your list of faults, a drawing of the resulting output for each of these faults, and an explanation of the output.

QUESTIONS/PROBLEMS

1. The output frequency from a Wien-bridge oscillator can be approximated using the same equation we used to calculate the cutoff frequency of a single-pole active filter. Find this equation in your textbook and calculate the value of f_{out} for the oscillator.

    $f_{out} = $ _____

2. Calculate the percent of error between your calculated value of f_{out} and the value measured in step 2 of the procedure.

    % of error = _____

    How would you account for this error?

3. Discuss, in your own words, what you observed in this exercise.

The Colpitts Oscillator

- To demonstrate the operation of the discrete Colpitts oscillator.
- To demonstrate the effects of changing values of L and C on the operating frequency of an LC oscillator.

DISCUSSION

The Colpitts oscillator is an LC oscillator that is capable of operating at a much higher frequency than a typical RC oscillator (such as the Wien-bridge oscillator). A discrete Colpitts oscillator is shown in Figure 52.1. The output frequency of the circuit is determined by L_T, C_1, and C_2, as will be shown in this exercise.

LAB PREPARATION

Review section 17.4 of *Introductory Electronic Devices and Circuits*.

LAB OVERVIEW

In this exercise, you will:

1. Construct a discrete Colpitts oscillator and measure its dc biasing voltages.
2. Observe the output waveform from the oscillator and measure its frequency.

3. Observe the effects of changing the lowest value capacitor on the output frequency.

MATERIALS

1 Variable dc power supply
1 VOM or DMM
1 Oscilloscope
1 2N3904 npn transistor
2 Resistors: 39 kΩ, 270 kΩ
2 Rf inductors: 1 mH, 10 mH
4 Capacitors: 51 pF, 100 pF, 0.01 µF, 0.022 µF
1 22 pF capacitor (additional-optional)

> Note: You will need to use an X10 oscilloscope probe in order to reduce circuit loading.

PROCEDURE

1. Construct the circuit shown in Figure 52.1.

> Note: When constructing the circuit, use short, direct leads. This will help prevent the circuit from picking up any undesired noise.

2. Apply power to the circuit. Measure and record the following:
   $V_B =$ _____
   $V_C =$ _____
   $V_B =$ _____

3. Observe the output waveform from the circuit. Draw this waveform in the space provided on the following page.

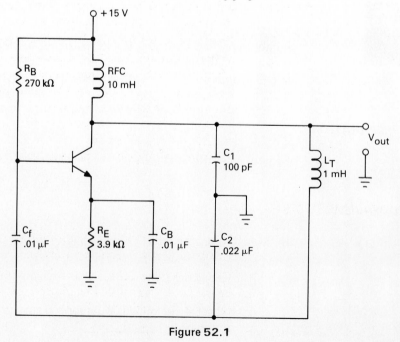

Figure 52.1

**Step 3
Waveform**

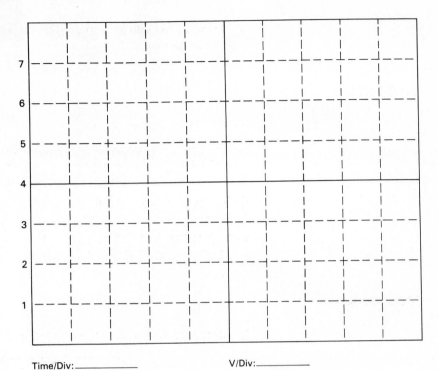

Time/Div:_____   V/Div:_____

4. Measure and record the oscillator output frequency and amplitude.

    f_{out} = _____ V_{out} = _____V_{PP}

5. Disconnect power from the circuit. Replace C_1 with your 51 pF capacitor. Reapply power and measure the circuit output frequency.

    f_{out} = _____

THE BRAIN DRAIN (Optional)

6. Return C_1 to its original value. Then, using your 22 pF capacitor, modify the circuit to form a Clapp oscillator.
7. Predict and measure the output frequency for the oscillator.
8. Verify, through circuit measurements, that the voltage gain of the circuit is approximately equal to the ratio of C_2 to C_1.
9. In a separate report, include the following:
    a. Your predicted and measured values of f_{out}
    b. The percent of error between the f_{out} values
    c. Your circuit schematic
    d. Your circuit gain verification procedure and results.

QUESTIONS/PROBLEMS

1. Using the equations shown in Figure 17.10 of the text, calculate the value of f_{out} for the Colpitts oscillator.

    f_{out} = _____

2. Calculate the percent of error between your calculated and measured values of f_{out} for the circuit.

    % of error = _____

How would you account for this error?

3. Compare the values of f_{out} measured in steps 4 and 5 of the procedure. Based on your measurements, what is the relationship between f_{out} and the capacitance in the feedback network?

4. Discuss, in your own words, what you observed in this exercise.

PART XI

Switching Circuits

Exercise 53

Switching Circuit Measurements

OBJECTIVES

- To provide the opportunity for measuring BJT switching times.
- To demonstrate the effects of a speed-up capacitor on BJT switching time.

DISCUSSION

The maximum switching rate of a BJT switch is limited by *delay time* (t_d), *rise time* (t_r), *storage time* (t_s), and *fall time* (t_f). Of the four, storage time is the primary contributor to the total propagation delay of the circuit.

Delay time and storage time can be drastically reduced by the use of a *speed-up capacitor*. This capacitor, placed in parallel with the circuit base resistor, provides a high initial reverse bias (reducing storage time) and a high initial forward value of I_B (reducing delay time).

LAB PREPARATION

Review sections 18.1 and 18.2 of *Introductory Electronic Devices and Circuits*.

LAB OVERVIEW

In this exercise, you will:

1. Measure the values of t_d, t_r, t_s, and t_f for a basic BJT switching circuit.

2. Add a speed-up capacitor to the circuit and observe the effect it has on the values of t_d, t_r, t_s, and t_f.

MATERIALS

1 Variable dc power supply
1 Variable square wave generator
1 Dual-trace oscilloscope
2 Resistors: 2.2 kΩ, 8.2 kΩ
1 100 pF capacitor

PROCEDURE

1. Construct the circuit shown in Figure 53.1.
2. Set your square wave generator for a 0V to +5 V output at a frequency of approximately 100 kHz.
3. Set your oscilloscope for dc coupling. Observe the input and output waveforms simultaneously. Draw these waveforms in the space provided.

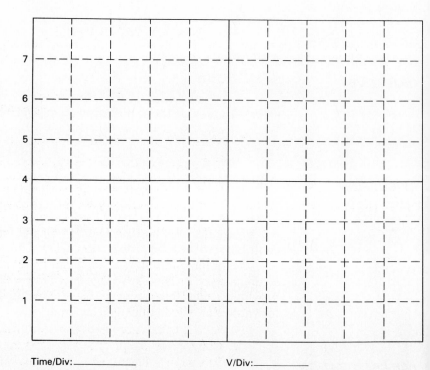

Time/Div:_____   V/Div:_____

4. By adjusting your TIME/DIV setting and the horizontal position of your oscilloscope traces, obtain a display like the one shown in Figure 53.2.

> Note: You can save some time and trouble by using the X5 setting in the oscilloscope time base section if it has such a setting.

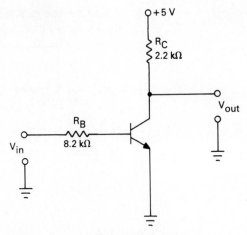

Figure 53.1

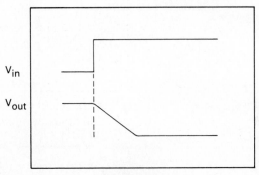

Figure 53.2

5. Measure and record the following values:

t_d = _____ t_r = _____

> Note: See Figure 18.23 in the text if you can't
> remember where the indicated times are measured
> on the waveforms.

6. Change the slope polarity control on the oscilloscope so that it triggers
   on the alternate transition. Then, adjust the horizontal position control
   until you see the trailing edge of the output waveform. Measure and
   record the following.

   t_s = _____ t_f = _____

7. Connect the 100 pF speed-up capacitor across R_B and then repeat steps
   3 through 5. Record your measured values in the spaces provided.

   t_d = _____ t_r = _____

   t_s = _____ t_f = _____

QUESTIONS/PROBLEMS

1. Of the values listed in steps 4 and 5, which was the longest? Do your
   values back the statement made regarding storage time in the discussion of this exercise?

2. Calculate the percent of change in each of your time measurements using

$$\% \text{ of change} = \frac{\text{initial time} - \text{new time}}{\text{initial time}} \times 100$$

% of change in t_d = _____

% of change in t_r = _____

% of change in t_s = _____

% of change in t_f = _____

Are these results consistent with the text explanation of the effects of a speed-up capacitor on BJT switching times? Explain your answer.

3. Using the values measured in step 6, calculate the theoretical and practical limits on f_{in} for the circuit used in this exercise.

f_{max} = _____   (theoretical)

f_{max} = _____   (practical)

4. Discuss, in your own words, what you observed in this exercise.

Op-Amp Schmitt Triggers

OBJECTIVES

- To demonstrate the operation of the inverting Schmitt trigger.
- To demonstrate the operation of the noninverting Schmitt trigger.

DISCUSSION

A Schmitt trigger is a *voltage-level detector*. That is, it provides a dc output that indicates when the input makes a positive-going transition past a given voltage called the *upper trigger point* (UTP) or a negative-going transition past a given voltage called the *lower trigger point* (LTP). The UTP and LTP voltage values may or may not be equal in magnitude, depending on the exact circuit configuration.

There are two types of Schmitt triggers. The *inverting* Schmitt trigger provides a *low* output (approximately equal to $-V + 1$ V) when the input passes the UTP, and a *high* output (approximately equal to $+V - 1$ V) when the input passes the LTP. The *noninverting* Schmitt trigger has the exact opposite input/output relationship.

LAB PREPARATION

Review section 18.3 of *Introductory Electronic Devices and Circuits*.

LAB OVERVIEW

In this exercise, you will:

1. Predict and measure the UTP and LTP values for an inverting Schmitt trigger.
2. Vary one of the resistor values in the inverting Schmitt trigger and observe the effect it has on the circuit output.
3. Predict and measure the UTP and LTP values for a noninverting Schmitt trigger.

MATERIALS

1   Dual-polarity variable dc power supply
1   Variable ac signal generator
1   Dual-trace oscilloscope
1   μA741 op-amp (or equivalent)
2   1N4148 small-signal diodes
4   Resistors: 1.1 kΩ, 2.2 kΩ, 3.3 kΩ, 11 kΩ

PROCEDURE

1. Construct the circuit shown in Figure 54.1.
2. Using equations (18.10) and (18.11) in the text, predict the UTP and LTP values for the circuit.

   UTP = _____   LTP = _____
3. Apply a 10 V_{PP}, 500 Hz input to the circuit.
4. Observe the input and output waveforms simultaneously. Draw these waveforms in the space provided on the following page.
5. Establish the center line on the CRT as the ground position for both traces.
6. With both traces having the same ground reference, the UTP and LTP values can be measured as shown in Figure 54.2. Measure these values.

   UTP = _____   LTP = _____

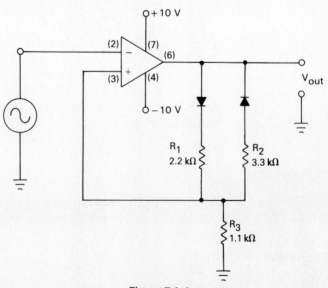

Figure 54.1

**Step 4
Waveforms**

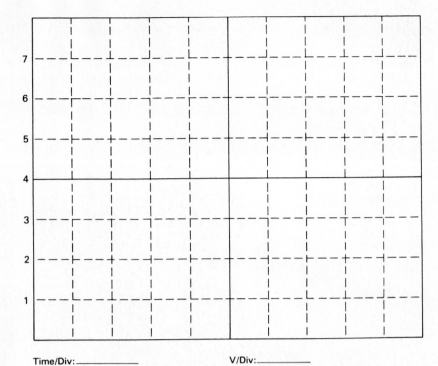

Time/Div:_____   V/Div:_____

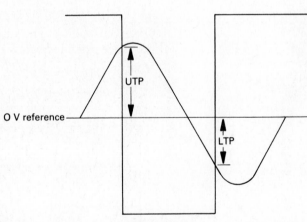

Figure 54.2

7. Disconnect power from the circuit and replace R_1 with the 11 kΩ resistor. Reapply power to the circuit and observe the input and output waveforms. Draw these waveforms in the space provided on the following page.

8. Measure and record the UTP and LTP values for the circuit.

   UTP = _____   LTP = _____

Part II. The Noninverting Schmitt Trigger

9. Construct the circuit shown in Figure 54.3.

10. Predict the UTP and LTP values for the circuit.

    UTP = _____   LTP = _____

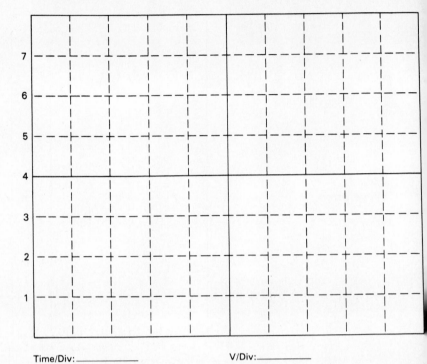

**Step 7
Waveforms**

Time/Div:_____   V/Div:_____

11. Apply a 10 V$_{PP}$, 500 Hz input to the circuit.
12. Observe the input and output waveforms. Draw these waveforms in the space provided below.
13. Measure and record the UTP and LTP values.

   UTP = _____   LTP = _____

**Step 12
Waveforms**

Time/Div:_____   V/Div:_____

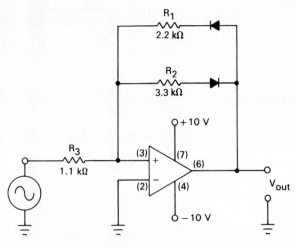

Figure 54.3

QUESTIONS/PROBLEMS

1. Calculate the percents of error between your predicted and measured UTP and LTP values (steps 2 and 6 of the procedure).

   % of error (UTP) = _____

   % of error (LTP) = _____

   How would you account for these errors?

2. In step 7 of the procedure, you changed the value of R_1. This caused the UTP of the circuit to change. What would have happened had you increased the value of R_2 instead of R_1? R_3?

3. Discuss, in your own words, what you observed in this exercise.

The 555 Timer One-Shot

OBJECTIVES

- To demonstrate the basic operation of the one-shot.
- To demonstrate the effects of changing circuit resistances on the output pulse duration of a one-shot.

DISCUSSION

The 555 timer contains all of the active components needed to construct either a *one-shot* (monostable multivibrator) or a *free-running* (astable) multivibrator. All you need to put together one of these circuits is a 555 timer and a few passive components.

The one-shot is a circuit that provides a single output pulse of predetermined duration when provided with an input trigger signal. A 555 timer one-shot is shown in Figure 55.1. The trigger signal is provided at pin 2 of the 555 timer. When triggered, the output from the timer (pin 3) goes high for a time that is determined by R_a and C_1. Then, the output automatically returns to its low voltage level. The output then remains low until another trigger signal is received.

LAB PREPARATION

Review section 18.4 of *Introductory Electronic Devices and Circuits*.

LAB OVERVIEW

In this exercise, you will:

1. Predict the output pulse duration for a 555 timer one-shot.
2. Take a series of trial measurements of the output pulse duration.
3. Change the value of R_a and observe the effect on the output pulse duration.

MATERIALS

1 Variable dc power supply
1 VOM or DMM
1 LM555 timer (or equivalent)
4 Resistors: 1 kΩ, 10 kΩ, 15MΩ, 22MΩ
2 Capacitors: 0.01 μF, 0.1 μF
1 Momentary push-button switch

PROCEDURE

1. Construct the circuit shown in Figure 55.1.
2. Measure and record the following voltages:
   V_T (pin 2) = _____
   V_{out} (pin 3) = _____
3. Using equation (18.13), predict the output pulse duration for the one-shot.
   PW = _____
4. Push the momentary switch to trigger the one-shot. When you trigger the circuit, you are going to estimate (as accurately as possible) the output pulse duration. (You will need a watch or stopwatch to do this.)

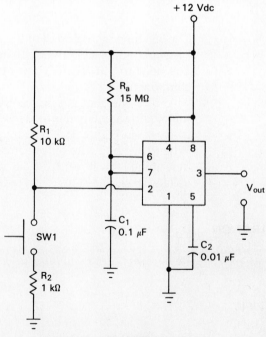

Figure 55.1

Perform this step three times, each time estimating the pulse width of the output pulse.

Trial 1: PW = _____

Trial 2: PW = _____

Trial 3: PW = _____

5. Calculate the average of the PW values listed in step 4.

PW = _____ (average)

6. Replace R_a with your 22MΩ resistor. Predict the output pulse duration.

PW = _____

7. Repeat step 4 for the circuit.

Trial 1: PW = _____

Trial 2: PW = _____

Trial 3: PW = _____

8. Calculate the average of the PW values listed in step 7.

PW = _____ (average)

QUESTIONS/PROBLEMS

1. Calculate the percent of error between the PW values obtained in steps 3 and 5 of the procedure.

% of error = _____

How would you account for this error?

2. What happened to the pulse duration of the circuit when you changed the value of R_a? Based on your answer, predict what would happen if you decreased the value of C_1.

3. Assume that you are designing a 555 timer one-shot for an output pulse duration of 15 ms. Using standard value components, list the values of R_a and C_1 you would use, and explain how you came up with those values.

4. Calculate the percent of error between the PW values you obtained in steps 6 and 8 of the procedure.
   % of error = _____

5. Compare the percents of error calculated in questions 1 and 4. How would you account for any difference between the two error values?

6. Discuss, in your own words, what you observed in this exercise.

The 555 Timer Free-Running Multivibrator

OBJECTIVE

- To demonstrate the operation of the free-running multivibrator.
- To demonstrate the effects of component changes on the output signal from a free-running multivibrator.

DISCUSSION

The 555 timer can be used as an *astable, or free-running, multivibrator*. When wired as such, the circuit will produce a constant square wave output. The free-running multivibrator is shown in Figure 56.1.

In this configuration, C_1 charges through R_a and R_b until the timer threshold voltage is reached. At that point, pin 7 drops to ground potential, and C_1 dis-

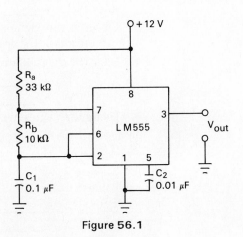

Figure 56.1

charges through R_b until a lower threshold voltage is reached. Then, pin 7 goes high again and the process repeats itself.

LAB PREPARATION

Review section 18.4 of *Introductory Electronic Devices and Circuits*.

LAB OVERVIEW

In this exercise, you will:

1. Predict and measure the cycle time and duty cycle of a free-running multivibrator.
2. Observe the effects of a change in R_a on the output cycle time and duty cycle of the circuit.
3. Observe the effects of a change in R_b on the output cycle time and duty cycle of the circuit.

MATERIALS

1 Variable dc power supply
1 Dual-trace oscilloscope
1 LM555 timer (or equivalent)
4 Resistors: 10 kΩ, 22 kΩ, 33 kΩ, 47 kΩ
2 Capacitors: 0.01 µF, 0.1 µF
1 10 kΩ potentiometer (optional)

PROCEDURE

1. Construct the circuit shown in Figure 56.1.
2. Using equations (18.18) and (18.19) in the text, predict the values of f_0 and duty cycle for the circuit.

   f_0 = _____

   duty cycle = _____

3. Observe the signals at pins 6 and 3 of the timer. Neatly draw the waveforms in the space provided.
4. Measure the following output values for the circuit.

   T_C = _____

   PW = _____

5. Using the values measured in step 4, calculate the following:

   f_0 = _____

   duty cycle = _____

6. Replace R_a with your 47 kΩ resistor. Measure and record the following values:

   T_C = _____

   PW = _____

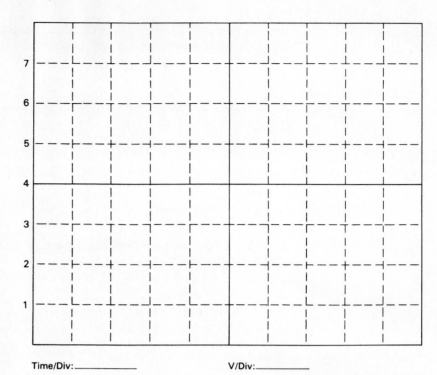

Time/Div:_____   V/Div:_____

7. Using the values measured in step 6, calculate the following:

   f_0 = _____

   duty cycle = _____

8. Replace R_b with your 22 kΩ resistor. Measure and record the following:

   T_C = _____

   PW   = _____

9. Using the values measured in step 8, calculate the following:

   f_0 = _____

   duty cycle = _____

THE BRAIN DRAIN (Optional)

10. Using your potentiometer, convert your free-running multivibrator into a voltage-controlled oscillator (as shown in Figure 18.52 in the text).

11. Use your circuit to verify the relationship between the circuit control voltage and f_0 (as stated in the text).

12. In a separate report, include your circuit schematic, a brief discussion on the relationship between control voltage and operating frequency, your testing method, and results.

QUESTIONS/PROBLEMS

1. Calculate the percent of error between your predicted and measured values of f_0 and duty cycle in steps 2 and 5 of the procedure.

   % of error in f_0 = _____

% of error in duty cycle = _____
How would you account for these errors?

2. What happened to the values of f_0 and duty cycle when R_a was changed?

3. What happened to the values of f_0 and duty cycle when you changed the value of R_b?

4. Compare the changes that occurred in steps 7 and 9 of the procedure. In your opinion, which change had the greatest effect on the value of f_0? On the duty cycle?

5. Discuss, in your own words, what you observed in this exercise.

Exercise 57

The Silicon Controlled Rectifier

OBJECTIVES

- To demonstrate the operating characteristics of the Silicon Controlled Rectifier (SCR).
- To demonstrate the operation of the SCR in dc and ac circuits.

DISCUSSION

SCRs are classified as *thyristors*. They are three-terminal, four-layer devices that use internal feedback to produce a *latching* (breakover) action. Introduced in 1956 by Bell Laboratories, SCRs are *unidirectional* devices. That is, they conduct in one direction only.

LAB PREPARATION

Review section 19.2 of *Introductory Electronic Devices and Circuits*.

LAB OVERVIEW

In this exercise, you will:

1. Construct an SCR test circuit and use that circuit to measure the values of I_{GT}, V_{GT}, V_{TM}, and I_H for the SCR.
2. Compare the measured values (in step 1 above) with the values listed on the SCR's specification sheet.
3. Construct a basic SCR phase-controller and observe its operation.

MATERIALS

1 Variable dc power supply
1 Variable ac signal generator
1 Dual-trace ocsilloscope
3 VOMs and/or DMMs
1 1N4001 rectifier diode
1 2N4444 SCR, along with its specification sheet.
1 LED
4 Resistors: 120 Ω, 510 Ω, 1 kΩ (2)
3 Potentiometers: 10 kΩ (2), 50 kΩ

PROCEDURE

1. Construct the circuit shown in Figure 57.1. The initial value of R_G should be 10 kΩ and the initial value of R_H should be 0 Ω.
2. Apply power to the circuit. Measure and record the following values:
   I_{AK} (the SCR anode current) = _____
   I_G (the SCR gate current) = _____
   V_{AK} (the SCR anode-to-cathode voltage) = _____
   The LED is _____.
             (on/off)
3. Slowly decrease the setting of R_G until the LED lights.

> Note: Stop adjusting R_G as soon as the LED lights. Do not continue to decrease its setting past the LED turn-on point.

4. Measure and record the following:
   I_{GT} (the SCR gate turn-on current) = _____

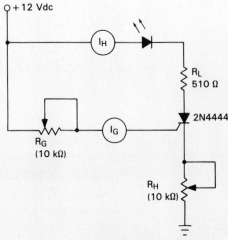

Figure 57.1

V_{GT} (the SCR gate turn-on voltage) = _____

V_{TM} (the SCR on-state voltage) = _____

> Note: V_{TM} is the on-state value of V_{AK}, measured across the SCR.

5. Increase the value of R_G to its maximum setting. Note the condition of the LED.

    The LED is _____ .
    (on/off)

6. Slowly increase the setting of R_H. Note the value of I_{AK} at the point just before the LED turns off. Record this value below.

    I_H (the SCR holding current) = _____

7. If you do not get a clear reading of I_H, retrigger the SCR and repeat step 6.

8. In the space provided below, record the values of I_{GT}, V_{GT}, V_{TM}, and I_H shown on the specification sheet of your SCR.

    I_{GT} = _____ V_{GT} = _____

    I_H = _____ V_{TM} = _____

9. Construct the circuit shown in Figure 57.2. R_1 should initially be set to approximately its midpoint position.

10. Apply a 12 V_{pk} input to the circuit at a frequency of approximately 1 kHz.

11. Observe the input and load waveforms simultaneously with your oscilloscope. While observing the output waveform, adjust R_1 to obtain the maximum peak load voltage.

12. Draw the circuit input and output waveforms in the space provided below.

Time/Div: _____   V/Div: _____

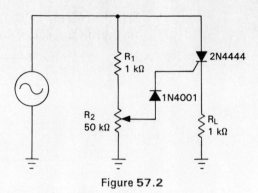

Figure 57.2

13. Vary the setting of R_1 and note the effect on the circuit output wave-form.

QUESTIONS/PROBLEMS

1. Compare the spec sheet values recorded in step 8 of the procedure with the measured values in steps 4 and 6. For each pair of values, calculate the percent of error between the rated and actual values.

% of error in I_{GT} = _____

% of error in I_H = _____

% of error in V_{GT} = _____

% of error in V_{TM} = _____

2. In step 5 of the procedure, why didn't the LED turn off when R_G increased to its maximum setting?

3. Refer to step 6 of the procedure. Why did the LED turn off when you increased the value of R_H?

4. Discuss, in your own words, what you observed in this exercise. (Use a separate sheet of paper, if necessary.)

The Unijunction Transistor

OBJECTIVES

- To demonstrate the characteristics of the unijunction transistor (UJT).
- To demonstrate the operation of the UJT as a simple relaxation oscillator.

DISCUSSION

The unijunction transistor (UJT) is a thyristor triggering device that is commonly used as the active element in a relaxation oscillator. The UJT is a three terminal device whose base 1-to-base 2 resistance rapidly decreases when the emitter-base 1 voltage reaches a specified value.

In this exercise, we will observe the operating characteristics of the UJT, as well as those of a basic relaxation oscillator.

LAB PREPARATION

Review section 19.4 of *Introductory Electronic Devices and Circuits*.

LAB PREPARATION

In this exercise, you will:

1. Build a UJT test circuit and observe its triggering characteristics.
2. Construct a basic relaxation oscillator and observe its operation.

MATERIALS

2   Variable dc power supplies
2   VOMs and/or DMMs
1   Dual-trace oscilloscope
1   2N4870 UJT, along with its specification sheet
2   Resistors: 1 kΩ and 51 kΩ
1   0.47 μF capacitor

PROCEDURE

1. Obtain the following ratings from the spec sheet for your 2N4870.

    $$\eta \ = \ \rule{3cm}{0.4pt}$$
    $$I_P \ = \ \rule{3cm}{0.4pt}$$
    $$R_{BB} \ = \ \rule{3cm}{0.4pt}$$
    $$V_{BB(max)} \ = \ \rule{3cm}{0.4pt}$$
    $$I_V \ = \ \rule{3cm}{0.4pt}$$
    $$V_{OB1} \ = \ \rule{3cm}{0.4pt}$$

    > Note: V_{BB} may be given as V_{B2B1} on your spec sheet.

2. Construct the circuit shown in Figure 58.1. The emitter supply voltage should initially be set to 0 Vdc and the emitter potentiometer should initially be set to approximately 50 kΩ.

3. Slowly increase the emitter supply voltage while measuring V_{EB1}. Continue to increase the supply voltage until V_{EB1} reaches a peak value. Measure and record the following:

    $$I_E \ = \ \rule{3cm}{0.4pt}$$
    $$V_{EB1} \ = \ \rule{3cm}{0.4pt}$$

    > Note: These are the peak current (I_P) and peak voltage (V_P) values for the UJT.

4. Vary the emitter supply voltage until V_{EB1} reaches a minimum value. Measure and record the following:

    $$I_E \ = \ \rule{3cm}{0.4pt}$$
    $$V_{EB1} \ = \ \rule{3cm}{0.4pt}$$

    > Note: These are the valley current (I_V) and valley voltage (V_V) values for the UJT.

5. Construct the circuit shown in Figure 58.2.

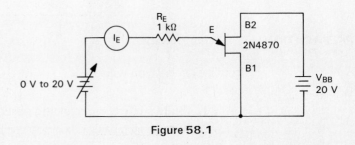

Figure 58.1

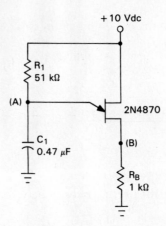

Figure 58.2

6.   Using your oscilloscope, observe the waveforms at points (A) and (B) in the circuit. Draw these waveforms in the space provided.

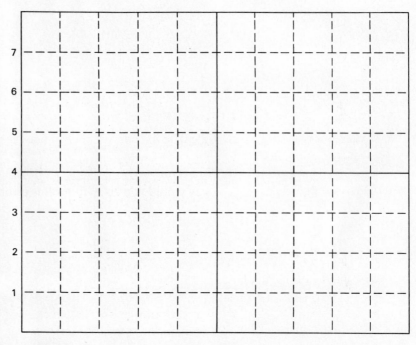

Time/Div:_____   V/Div:_____

7.   Measure and record the following values:

V_{pk} (point A) = _____

Cycle time = _____

Note: V_{pk} at point A is the value of V_{OB1} for your UJT.

8.   Calculate the value η for your UJT using the following equation:

$$\eta = \frac{V_{OB1} - 0.7V}{V_{BB}}$$

η = _____

9. Calculate the cycle time of the relaxation oscillator using the following equation:

$$T_C = R_1 C_1 \left[\ln \left(\frac{1}{1 - \eta} \right) \right]$$

$T_C =$ _____

QUESTIONS/PROBLEMS

1. Calculate the percent of error between your measured and rated values of I_V.
    % of error = _____
    How would you account for this error?

2. Calculate the percent of error between your calculated and rated values of η.
    % of error = _____
    How would you account for this error?

3. The negative resistance region of operation is defined (limited) by the values found in steps (3) and (4) of the procedure. Based on your measurements, is the term negative resistance accurate? Explain your answer.

4. Calculate the percent of error between your measured and calculated values of T_C for the relaxation oscillator.

% of error = _____

How would you account for this error?

5. Discuss, in your own words, what you observed in this exercise. (Use a separate sheet of paper, if necessary.)

Exercise 59

Opto-Isolators

OBJECTIVE

- To demonstrate the operating characteristics of discrete and integrated opto-isolators.

DISCUSSION

The opto-isolator, or opto-coupler, is simply a package that contains both an infrared LED (IRED) and a photo-detector (such as a photo-diode, photo-transistor, etc.). The greatest advantage of using this component is that the IRED and detector are completely isolated from each other, yet signals can be readily transferred between the two. This makes it possible for low voltage and current output devices to control high voltage and current demanding loads with complete isolation between the source and load circuits.

Opto-isolators are rapidly replacing mechanical relays, isolation transformers and coupling capacitors in many applications. The reason for this is the fact that opto-isolators have high-frequency switching capabilities, low losses, high reliability, and relatively small size.

LAB PREPARATION

Review section 19.7 of *Introductory Electronic Devices and Circuits*.

LAB OVERVIEW

In this exercise, you will:

1. Build and analyze a discrete opto-isolator.
2. Measure the frequency response of the discrete opto-isolator.
3. Observe the operation of an integrated opto-isolator.
4. Measure the frequency response of the integrated opto-isolator.

MATERIALS

1 Variable dc power supply
1 Variable ac signal generator
1 VOM and/or DMM
1 Dual-trace oscilloscope
1 4N35 opto-isolator (transistor output)
1 MLED71 Infrared LED
1 MRD300 Photo-transistor
3 Resistors: 220 Ω, 820 Ω, and 1 kΩ
2 Potentiometers: 1 kΩ and 10 kΩ
2 Capacitors: 1 μF and 47 μF
1 2.5 cm (1 inch) piece of heat shrink tubing

PROCEDURE

1. Construct the circuit shown in Figure 59.1. Use the heat shrink tubing
   to connect the IRED to the photo-transistor to prevent interference
   from outside light sources. R_1 should initially be set to approximately
   700 Ω, and R_2 should initially be set to approximately 5 kΩ.

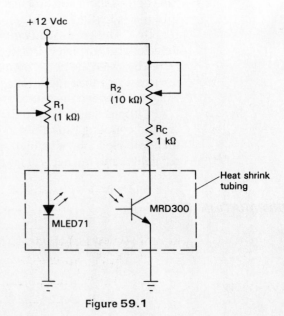

Figure 59.1

2. Apply power to the circuit and adjust R_1 so that V_C of the transistor is approximately 6 Vdc.

3. Measure and record the following:

$$V_C \ = \ \underline{\hspace{3cm}}$$

$$I_C \ = \ \underline{\hspace{3cm}}$$

I_F (of the diode) $= \ \underline{\hspace{3cm}}$

4. Using the values of I_C and I_F, calculate the current-transfer ratio ($I_C{:}I_F$) of the opto-isolator.

$$CTR \ = \ \underline{\hspace{3cm}}$$

5. Apply a 1 kHz input from the ac signal generator. Adjust the input amplitude to provide the maximum undistorted output from the photo-transistor.

6. Draw the input and output waveforms for the opto-isolator in the space

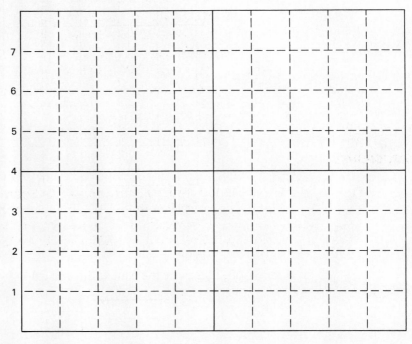

Time/Div:_____   V/Div:_____

provided.

7. Using the techniques outlined in Exercise 41, measure the bandwidth and cutoff frequencies of the opto-isolator.

$$f_1 \ = \ \underline{\hspace{3cm}}$$

$$f_2 \ = \ \underline{\hspace{3cm}}$$

$$BW \ = \ \underline{\hspace{3cm}}$$

8. Construct the circuit shown in Figure 59.2.

9. Apply power to the circuit. Measure and record the following values:

$$V_C \ = \ \underline{\hspace{3cm}}$$

$$I_C \ = \ \underline{\hspace{3cm}}$$

$$I_F = \underline{\hspace{2cm}}$$

10. Calculate the current transfer ratio of the 4N35.

    $$CTR = \underline{\hspace{2cm}}$$

11. Apply a 1 kHz input to the circuit. Adjust the input amplitude for the maximum undistorted output signal.

12. In the space provided, draw the input and output waveforms for the

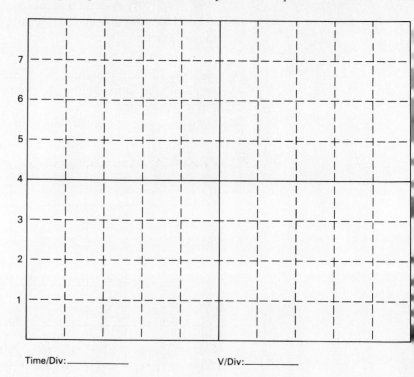

Time/Div:_____   V/Div:_____

circuit.

13. Measure the bandwidth and cutoff frequencies of the 4N35.

    $$f_1 = \underline{\hspace{2cm}}$$

    $$f_2 = \underline{\hspace{2cm}}$$

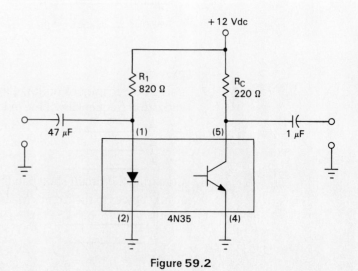

Figure 59.2

BW = _____

QUESTIONS/PROBLEMS

1. Based on your results in this exercise, what are the advantages of using integrated opto-isolators over discrete opto-isolators?

2. How does the bandwidth of the integrated circuit compare with that of the discrete circuit?

3. Discuss, in your own words, what you observed in this exercise. (Use a separate sheet of paper, if necessary.)

Exercise 60

Discrete Voltage Regulators

OBJECTIVES

- To demonstrate the basic operation of a pass-transistor voltage regulator.
- To demonstrate the use of a current limiting circuit.

DISCUSSION

The pass-transistor regulator is a series voltage regulator. The basic pass-transistor regulator, shown in Figure 60.1 uses a series (or pass) transistor to adjust the circuit output current as the load demand changes.

A current-limiting transistor can be added to the basic pass-transistor regulator in order to protect the pass-transistor from excessive load current demands. Such a protection circuit is shown in Figure 60.2.

LAB PREPARATION

Review sections 20.1 and 20.2 of *Introductory Electronic Devices and Circuits*.

LAB OVERVIEW

In this exercise, you will:

1. Construct the pass-transistor regulator and observe the reaction of the circuit to changes in load current demand.

2. Add a current-limiting circuit to the regulator and measure the output short-circuit current.

MATERIALS

2   VOMs and /or DMMs
1   24 Vac center-tapped transformer
2   1N4001 rectifier diodes
1   1N5240 zener diode
2   2N3904 npn transistors
4   Resistors: 33 Ω, 1 kΩ, 1.1 kΩ, 2.2 kΩ
2   Capacitors: 0.1 µF, 2200 µF

PROCEDURE

1. Construct the circuit shown in Figure 60.1. Do not apply power to the circuit until directed to do so.
2. Predict the following values for the circuit:

   V_{out} = _____

   I_{out} = _____
3. Apply power to the circuit. Measure and record the following values:

   V_{out} = _____

   I_{out} = _____
4. Disconnect power from the circuit. Replace R_L with your 1.1 kΩ resistor.
5. Reapply power to the circuit. Measure and record the following values:

   V_{out} = _____

   I_{out} = _____
6. Disconnect power from the circuit. Replace R_L with your DMM in order to measure the "no-load" output voltage.
7. Reapply power to the circuit. Measure and record the following values:

   V_{out} = _____

   I_{out} = _____
8. Modify your voltage regulator as shown in Figure 60.2. Do not apply power to the circuit until directed to do so.

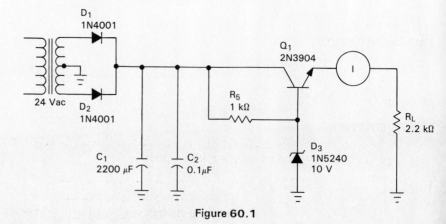

Figure 60.1

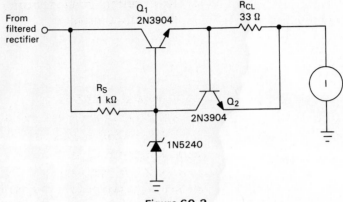

Figure 60.2

9. Predict the output short-circuit current using equation (20.5) from the text.

$I_{L(max)}$ = _____

10. Apply power to the circuit. Measure and record the value of load current.

$I_{L(max)}$ = _____

QUESTIONS/PROBLEMS

1. Calculate the percents of error between your predicted and measured values in steps 2 and 3 of the procedure.

% of error of error in V_{out} = _____

% of error in I_{out} = _____

How would you account for this error?

2. Using the values measured in steps 5 and 7 of the procedure, calculate the following values:

ΔV_{out} = _____

ΔI_{out} = _____

Load regulation = _____ V/mA

3. How did the voltage regulator respond to an increase in load current demand?

4. Calculate the percent of error between your predicted and measured values of $I_{L(max)}$ in steps 9 and 10 of the procedure.

   % of error = _____

   How would you account for this error?

5. Discuss, in your own words, what you observed in this exercise.

Exercise 61

IC Voltage Regulators

OBJECTIVES

- To demonstrate the operation of the three terminal IC voltage regulator.
- To demonstrate the load regulation capabilities of an IC voltage regulator.

DISCUSSION

The IC voltage regulator replaces the discrete voltage regulator in most practical power supply applications. This is due to the improved line regulation, load regulation, and power dissipation capabilities of the IC voltage regulator.

LAB PREPARATION

Review section 20.4 of *Introductory Electronic Devices and Circuits*.

LAB OVERVIEW

In this exercise, you will:

1. Construct a basic IC regulated dc power supply.
2. Measure the regulator's response to a change in load current demand and calculate its load regulation.

MATERIALS

2 VOMs and/or DMMs
1 24 Vac center-tapped transformer
2 1N4001 rectifier diodes
1 LM317 voltage regulator
4 Resistors: 1.1 Ω, 1.5 Ω, 2.2 kΩ, 8.2 kΩ
3 Capacitors: 0.1 μF, 1000 μF, 2200 μF

PROCEDURE

1. Construct the circuit shown in Figure 61.1. Do not apply power until directed to do so.
2. Using equation (20.7) in the text, predict the values of V_{out} and I_{out} for the circuit:
   V_{out} = _____ I_{out} = _____
3. Apply power to the circuit. Measure and record the following values:
   V_{out} = _____ I_{out} = _____
3. Disconnect power from the circuit and remove the load resistor. Replace the load resistor with your DMM so that you can measure the "no-load" output voltage for the circuit.
4. Reapply power to the circuit. Measure and record the following values:
   V_{out} = _____ I_{out} = _____
5. Disconnect power from the circuit. Replace R_L with your 1.1 k resistor.
6. Apply power to the circuit. Measure and record the following values:
   V_{out} = _____ I_{out} = _____

QUESTIONS/PROBLEMS

1. Calculate the percent of error between the predicted and measured values in steps 2 and 3 of the procedure.
   % of error in V_{out} = _____
   % of error in I_{out} = _____

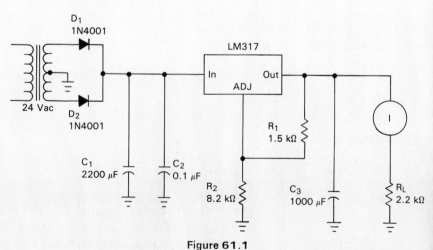

Figure 61.1

How would you account for these errors?

2. Using the values measured in steps 4 and 6 of the procedure, calculate the following values for the circuit:

ΔV_{out} = _____

ΔI_{out} = _____

Load regulation = _____V/mA

3. How did the circuit respond to an increase in load current demand?

4. Discuss, in your own words, what you observed in this exercise.

LAB REPORT REQUIREMENTS

A) DATA SHEETS

B) NEAT GRAPHS (LABELED AXES, ETC)

C) COVER SHEET

D) ANSWERS TO QUESTIONS (UNLESS OTHERWISE ADVISED | GREATER % OF GRADE)

E) SUPPLEMENTARY QUESTIONS

3 → 1

5   2

7   3

9   4

{4, 10}   5